W9-BHM-962

BIRDS AND THEIR YOUNG

◆

Gordon Dee Alcorn

Illustrations by Michelle LaGory

Courtship, Nesting, Hatching, Fledging
The Reproductive Cycle

Stackpole Books

Published by

STACKPOLE BOOKS
Cameron and Kelker Streets
P.O. Box 1831
Harrisburg, PA 17105

Printed in the United States of America

10 9 8 7 6 5 4 3 2 1

First Edition

Book design, typography, and production by Art Unlimited

Library of Congress Cataloging-in-Publication Data

Alcorn, Gordon Dee
Birds and their young: courtship, nesting, hatching, fledging—the reproductive cycle / Gordon D. Alcorn. — 1st ed.
p. cm.
Includes bibliographical references and index.
ISBN 0-8117-1016-5: $18.95
1. Birds—Reproduction. 2. Birds—Life cycles. 3. Birds—Infancy. 4 Birds—Behavior. I. Title.
QL698.2.A34 1991
598—dc20 90-39204
CIP

To my great-grandson Austin

With the hope that
someday he will share my love for the birds

CONTENTS

ACKNOWLEDGMENTS

It is a pleasure to acknowledge my debt to the people who worked long and diligently to make this book possible. Sandra Bauer, Eileen McDowell, Stephani Marsh, and Claudia Mekins typed and retyped many pages of the manuscript. Bruce Smith, Richard Fitzner, Robert Pyle, and Jack Cowan read and offered valuable suggestions that improved the book. Special thanks are due Susan McMahon and Toni Peterson, who worked long hours on all aspects of the manuscript. I greatly appreciate Manfred Hoberg's information about the hatching of the domestic chicken, and I am grateful to Fors Farms, Inc., for its gift of fertile eggs. To my wife Rowena I extend special thanks for her artwork, her consistent interest, and her encouragement for the project.

I am also very grateful to the following publishers who gave permission to quote from various publications in their inventories. (The chapter number in parentheses refers to the location of the quoted material in this book; a complete reference is provided in the bibliography.) Thanks go to the following: Harvard Press, for permission to borrow a most appropriate sentence from Wilson's *Sociobiology, the New Synthesis*: "There is a waiting list for prime real estate" (Chapter 4); The William Collins Co., for permission to print the list of artifacts collected by the Satin Bower Bird (Chapter 7); Doubleday and Co., for permission to use Gilliard's nest census of the weaver bird in *Living Birds of the World* (Chapter 5); the editor of *The Auk* for permission to use Friedmann's *et al.* references to cowbird eggs and discussion of rejector and acceptor hosts (Chapter 9), and for use of Mayfield's statements concerning the incidence of cowbird parasitism on Kirtland's warbler (Chapter 9); the Cornell University Press, for George Clark Jr.'s information on temperature testing by megapodes in *The Living Bird* (Chapter 10); *The Auk*, for an explanation of the development of the egg tooth and hatching muscle by H. I. Fisher (Chapter 11); the Wildfowl Trust, for permission to quote from L. H. Brown on the behavior in the creche (Chapter 12); and the editor of *The Condor* for Cade and MacLean's information on the behavior of the sand grouse with respect to drinking water (Chapter 12), Koenig's discussion of prey dunking by Brewer's blackbird (Chapter 12), and Weatherhead's observations on disposal of fecal sacs in birds (Chapter 12).

INTRODUCTION

The reproductive cycle in birds and the associated sciences of oology (eggs) and nidology (nests) are complex. From the initiation of the cycle and the enlargement of the gonads, each internal event parallels changes in the behavior of the parent birds and eventually the young as well, first when they leave the egg and later when they fledge. In most birds, the cycle follows the same general pattern with specific adaptations due to environmental forces, meteorological changes, and genetic influences. Birds must in a comparatively short time seize upon a habitat within a selected environment, called territory, and get all the preliminaries for a successful nesting cycle (courtship, mating, nest construction) out of the way. Once these preliminaries have been accomplished the parent birds can move into incubation, which is a more or less monotonous period of greatly reduced activity. After the eggs hatch the pace picks up a bit with caring for the young, but not to the same level as the frantic lifestyle necessary before egg laying.

To bring to print the details of the reproductive cycle from beginning to end for all bird species would be a monumental task. While the general cycle follows a fairly common pattern in all birds at all stages, there are modifications in almost every one of the world's approximately 8,500 species. These adaptations can be seen as attempts to overcome the difficulties or take advantage of the assets of an environment, thereby making the fledgling stage more successful.

The following pages outline the general steps that parent birds go through to succeed in the fledging of their young. I have tried to use sufficient examples to point out the overall similarities among different species. At the same time, I have tried to show how the behaviors of some birds change in unusual conditions such as extreme heat or cold. Such extremes make reproductive processes more difficult, and certain species of birds must adapt in one way or another to avoid disastrous results.

For most species of birds, a significant part of their year—up to ten or eleven months in the megapodes—is devoted to reproductive activity. Instincts for reproduction are highly developed in the thousands of species of birds, and because of these special instincts birds show a marvelous and colorful series of behavioral, structural, and physiological patterns to establish the reproductive cycle and raise their young.

A few birds, such as brood parasites, modify their behavior to exclude some of the normal steps, such as incubation and feeding, instead assigning those tasks to host parents. Consequently, for expediency, certain behavioral activities must be substituted for others. Parasitic females must, for example, instinctively know how to select a suitable host nest and when to lay their eggs there.

In the discussion here, repeated references are made to various historical items. References to old egg records that are scattered throughout the text are taken from original data blanks for egg sets found in the Slater Museum at the University of Puget Sound Museum of Natural History. These are referred to in the text as UPS Museum. Many of these sets with accompanying records date from the late 1800s and the early 1900s.

Oology as a science reached its peak at about the turn of the century. At this time egg collecting was extremely popular among scientists, amateurs, and curious bird lovers of all ages. But with the coming of state and federal laws prohibiting the indiscriminate collecting of bird eggs, the number of collectors declined. For a time, states issued collecting permits for interested private collectors. But today most collecting permits are issued only to scientists associated with institutions. Not surprisingly, the rise of protectionism was paralleled by a sharp decline in oology and nidology; the general population today has little interest in these fields. Most large collections are managed by government agencies or colleges and universities in their museums.

Despite the fact that the people are prohibited by law from an active program of research and collection, the aesthetics of bird study are highly developed in the general population. Museum displays of bird nests and eggs are just as popular as they have ever been. Oology and nidology appeal to the average person because the colors, patterns, and mechanics of eggs and nest building arouse the artistic interest inherent, to some degree, in most people. R. Magoon Barnes said it well:

> "Perhaps the most marvelous of nature's mysteries relate to the reproduction of animal life. And one of the methods employed by nature, is to wrap the life spark of the family of Aves (Birds) up within a hard, partially porous shell composed largely of lime, with a generous supply of food for the young bird; and to cause this life spark to germinate during a period of incubation, which varies in length with different birds, during which in most cases, the egg is kept warm by the mother bird sitting on it. At the end of this period of incubation the young bird hatches, that is it breaks out of the shell. The size, shape, color markings and texture of the shells of birds' eggs is of infinite variety, and their beauty is proverbial."

THE COMPLEXITY OF THE REPRODUCTIVE CYCLE

The reproductive cycle in birds follows an observable sequence, beginning with various rituals of courtship and ending with a fledgling bird leading an independent life. Courtship and selection of territory are followed by nest building, egg laying, incubation, hatching, and care and training of young. For reproduction to succeed the bird must take advantage of all the environmental factors that affect the cycle and must use its instincts to circumvent any threats to the eggs and young.

Where there is food and some sort of shelter in the world, birds will be found living in harmony with the given environment. But they may not be compatible with their surroundings in all seasons of the year. Thus the phenomenon of migration, which we see in many birds, is often necessary for the successful completion of the reproductive cycle. Migratory habits, which are partly due to physiological changes in the bird's body, guide many species into regions of the world that are highly suitable for reproduction in one season of the year but extremely hostile in another season. Great flocks of shore birds, waterfowl, and passerines, for example, annually move between temperate and northern latitudes. These northern regions, during the summer nesting season, provide long days, abundant food, and space. Huge numbers of these birds move rapidly north during the breeding cycle in the major flyways of North America, South America, Europe, Asia, and Africa. In spring an arctic island may be thickly popu-

lated with as many as 150,000 thick-billed murres, 150,000 fulmars, 100,000 glaucous gulls, and 100,000 black-legged kittiwakes. Large numbers of these birds nest side by side on cliffs—so many, in fact, that sometimes great numbers of kittiwakes, unable to find a nest site, simply sit out the breeding season.

Migratory flights to the north are direct and rapid, possibly because the northern season is so brief. On the return flights the energies and hormonal drives related to the reproductive cycle have lessened or ceased altogether. These flights are more leisurely and often less direct.

Attrition rates are high in many species, such as those that migrate long distances to find an environment with adequate space, nesting materials, and food for successful rearing of the young. Wind, rain, and extreme temperatures take their toll in the travel to and from nesting grounds; many birds die en route. Once at the destination, the nesting period is short. Still, seabirds, shore birds, and many passerines fly these long distances to spend the reproductive season in the north.

Latitudinal differences in the Southern Hemisphere have fewer influences on large populations of reproducing birds than do those in the Northern Hemisphere. Land masses in the Southern Hemisphere are small compared with the vast land masses and greater food availability in the continental reaches north of the equator. Tundras, for example, that are circumpolar in the Nearctic and Palearctic continents of the North are comparatively scarce south of the equator, but birds are successful nesters in extremes of both hemispheres. In the perpetual cold of the antarctic, large populations of penguins have developed various unusual adaptations that make reproduction not only possible, but successful. Breeding populations of the great skua (*Catharacta skua*) can be found in the arctic region of Iceland and also in Antarctica. (This skua has the distinction of being the only species that nests in both the arctic and the antarctic.)

The regions of the world richest in bird life are the Neotropical, and they are mostly distributed immediately north and south of the equator. Here the climate exhibits fewer drastic changes annually; nesting cycles are not subject to extremes of temperature or major differences in lengths of days.

Most birds, living where the seasons are marked, start their incubation periods during the most favorable time of the year, usually spring. Some birds, however, such as northern owls and some gulls, begin their incubation cycles in late winter while the land is still cold and there is snow on the ground. This insures that fledging occurs at the most favorable time of the year when temperatures are warm and food is plentiful. Parmelee records the acquisition of a set of fresh eggs of Thayer's gull (*Larus thayeri*) on June 21, 1960, at Finlayson Island, which is in the Northwest Territories near Victoria Island (UPS Museum). Parmelee noted that ice locked the shore that day; it was possible to reach the colony only by dogsled. The penguin on the opposite side of the world starts its reproductive cycle even earlier, at the beginning of winter. After two months of incubation, the young finally hatch at the peak of the antarctic summer.

What motivates birds to begin the reproductive portion of their life cycle and to bring it to a successful conclusion? This question has never been fully answered, but there are probably several factors involved. The chief of these may be the hormonal changes in the bird's body, which go through a yearly cycle in the adult bird. Perhaps length of day, diet, and temperature range also contribute.

Migration seems related to the reproductive cycle as well. When the reproducing cycle develops toward its peak, when hormones are at their maximum, and when the gonads are enlarged, birds tend to migrate. Which of these factors is the cause and which the effect is difficult to determine.

Behavior patterns are also affected by the hormonal cycle. At a predictable time, birds that were normally gregarious become solitary or perhaps paired. Birds that were typically even-tempered in the gregarious periods of their lives become aggressive. The types of food and the way they eat it may change, as will song, especially during the vocal portions of a courtship when songs may be more frequent and complex. But later, when the nest is built and the eggs are laid, these same birds become notably silent. The voices of many altricial birds, when helpless and naked young are in nests, become very strident, as if the adults were displaying anger or fear at the transgression of an interloper.

Reproductive success—measured by the perpetuation of generations—depends on the efficiency of a remarkable series of events. Birds and their ancestors, the reptiles, have mostly forsaken an aquatic life for a terrestrial one. They have met great success because they have developed a series of adaptations that are suitable for life on the land. One such adaptation is the hard, calcareous eggshell, which prevents the embryo from drying up and supports the interior soft parts of the egg and the developing young. Additionally, birds guard their eggs carefully in protective nests, whereas aquatic invertebrates, such as oysters, lay several thousand eggs and abandon them to the vagaries of wind and wave, thereby gambling that a half-dozen individuals will develop. Birds, with their terrestrial life and the consequent complexity of their reproductive cycle, have reduced the number of eggs laid per individual, but at the same time have protected these eggs and the young in a number of ways. The oyster that lays many thousands of eggs is no more successful in perpetuating its kind than the bird that lays but a few eggs.

Not only do birds protect their eggs, they take steps to insure the survival of the hatchlings. A hen leads her chicks around the barnyard, summons them at the discovery of food, and teaches them how to scratch for their dinner. A mother mountain quail (*Oreortyx pictus*), with a few chirping sounds, sends her brood scampering to hide in the grass until some danger passes. When the coast is clear, the mother appears and calls her brood back to her supervision. The barnyard hen uses a different tactic: she calls her young chicks to hide under her outspread, protective wings.

Such nurturing activity may be partially instinctive and partially learned, though it is difficult to separate behavior patterns into instinctive and learned components. A parent swallow flying past a group of young swallows lined up along a wire is obviously teaching her young brood to take food in motion, leading up to the day when the young must capture their own food on the wing. But is a duck fresh out of the shell taught to swim, or is the swimming instinctive? What prompts the downy young of the grebe and loon to climb up onto the backs of swimming adults, or directs a parent bird to feign a broken wing to lure a predator away from vulnerable eggs or young? These and many other activities probably lie somewhere between completely learned and completely instinctive behavior, but all contribute to the success of the reproductive cycle.

BEYOND REPTILES

THE

ORIGIN OF THE BIRD

There is general agreement among scientists that birds originated from primitive reptiles. Anatomical similarities between fossil birds and fossil reptiles as well as similarities between modern forms support this conclusion.

One of paleontology's most famous animal fossils is *Archaeopteryx lithographica*, found in 1861 in Bavaria in a Jurassic limestone quarry. Estimates put *Archaeopteryx's* age at 140 million years. The bone casts show distinctive reptilian traits but also show unmistakable avian characteristics. *Archaeopteryx* has teeth, a bony tail, a reptile-like skull, and claws on its wings. (Some modern birds possess vestigial claws on the wings. In the hoatzin (*Opisthocomos cristatus*), the claws make it possible for the fledgling to climb trees.) However, *Archaeopteryx* also shows feathers that are almost identical to the feathers on a modern bird. Additionally, there is some evidence of hollow bones, an avian characteristic. Consequently, insofar as ancient fossils are concerned, the *Archaeopteryx* is not a true "missing link" between primitive reptiles and their derivative, the bird, because there is little evidence linking structural characteristics of fossil reptiles of the earlier Triassic period to the bones of *Archaeopteryx*. Furthermore, there is no fossil evidence to show a step-by-step development of the modern feather form or type, as found in *Archaeopteryx*.

Neither is there clear-cut fossil evidence showing any transitional, intermediate structures between primitive reptilian forms and modern birds retaining unmis-

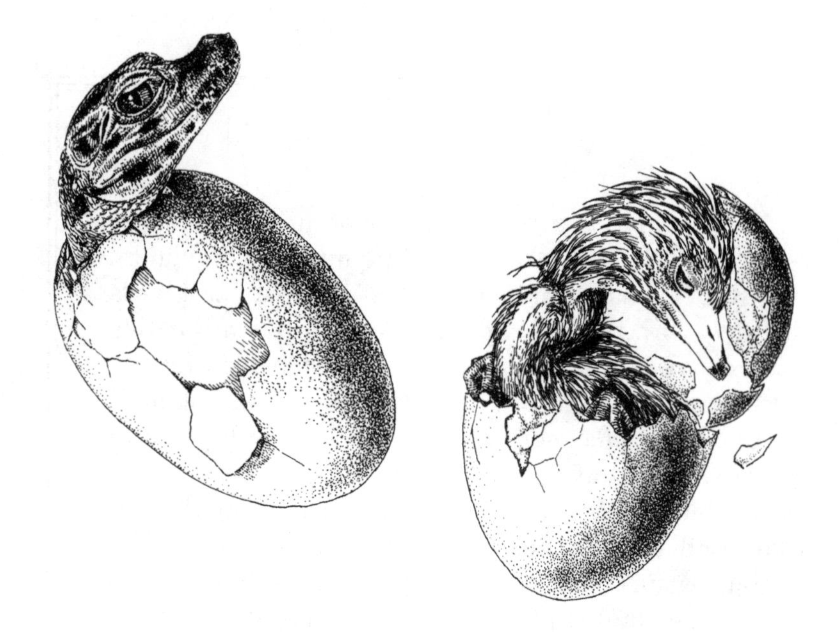

takably primitive, reptilian traits. Except for the presence of feathers in *Archaeopteryx,* this fossil could be classified as reptilian. Indeed, for a long time several specimens of *Archaeopteryx* were thought to be primitive reptiles.

Ancient fossils of birds and reptiles show striking similarities in skeletal parts and other structures. There are resemblances between bird and reptile skulls, neck vertebrae, ribs, various pneumatic bones, brains, eyes, blood, and eggs. Both groups also possess scales.

Our attention, however, should be focused on the reproductive structures and cycles of the two groups. In order to arrive at some conclusions regarding relationships between birds and reptiles it is valuable and indeed essential to briefly examine the physical, morphological, physiological, and behavioral patterns in all phases of the reproductive cycle in birds compared to those in reptiles.

Reptiles and birds lay similar eggs, and their young, when hatching from the shell, possess an egg tooth. In both reptiles and birds the egg is constructed to survive outside of water, as distinct from lower animals, such as amphibians. Water is essential to the eggs of amphibians: it provides oxygen and protects them from dessication.

This water stage is entirely missing in reptiles and birds.

Inside the egg of both reptile and bird is a yolk containing food in a yolk sac intimately associated in the growing embryo with the digestive system. On about the third day of incubation, a second and larger sac develops around the body of the embryo; this thin, transparent membrane is called the amnion. The liquid inside this membrane protects the embryo from dessication and mechanical injury.

A third structure, the allantois, develops in both reptile and bird eggs to serve, in part, as a temporary bladder that holds the protein waste of the developing embryo. Because of its function, the allantois grows from the posterior end of the intestinal tract and lies outside the amnion and immediately under the shell membrane or chorion. Between the chorion and the eggshell, oxygen and carbon dioxide filter through the shell and into and out of the allantois. In this position the allantois can serve as both excretory and respiratory systems for the embryo. (There is no urinary bladder in adult birds. Such a structure would be highly impractical because aerodynamic principles require that weight be kept to a minimum. Birds could not afford to develop a urinary bladder to temporarily store liquid urine. Rather than dispose of urea in liquid form, birds rid the body of protein waste as uric acid, which is a solid.) The allantois is the only structure capable of ridding the embryonic body of its uric acid waste; this substance is insoluble and would soon poison the embryo if allowed to accumulate throughout its body. The allantois thus becomes a temporary but essential part of the embryonic metabolic processes.

In both reptiles and birds a hard, protective shell develops around the entire egg—a calcareous cover for the soft and liquid parts of the inner egg. In comparing the eggs of reptiles with the eggs of birds it becomes obvious that the development of the protected egg allowed the two groups to live very successfully on land for the first time in history. The establishment of the amnionic type of egg in birds and reptiles is probably the most remarkable adaptation in vertebrate history. This move to land eliminated the need for the amphibian's larval stage; newly hatched birds and reptiles are immature forms of the adult.

Reptiles and birds also show some similarities in behavior patterns directly or even indirectly related to incubation. Some serpents demonstrate broodiness. The egg-laying snake may coil herself around a clutch of eggs, not so much, perhaps, for a transfer of heat as for egg protection. Some reptiles and birds also exhibit similar behavior when placing eggs, especially underground. Certain turtles commonly dig a hole in sand, place their eggs inside, cover them with sand, and depart—leaving the young, when hatched, to dig out for themselves. The alligator may place its eggs in a hole but stations herself nearby to prevent some marauding predator from digging out the eggs for food. In birds, some megapodes cover their eggs in sand and depart. Other megapodes (brush turkeys and mallee fowls) place their eggs in a mound of duff on the forest floor. These birds, however, do not leave the area but instead manipulate the mound to control its temperature throughout the incubation cycle, as discussed in Chapter 10. The gavial of India (*Gavialis gangeticus*), a large reptile found in the basins of the Ganges, Brahmaputra, and Indus rivers, shows a highly developed concern for its young. The female digs a hole in the sand of a river bank and lays several hundred eggs, which she very carefully covers with a couple feet of sand. This process may take as long as ten hours. She then stays nearby and guards the nest site. After seventy days of incubation the female digs out the young after locating the nest by listening for their chirping as they break free from the shell. Hatching may take a total of ten hours. A well-developed egg tooth appears on the snout of the young to aid in breaking free from the shell. While hatching proceeds the male comes to watch but the female makes him keep his distance. Curiously, other females—mostly immature—also come to the hatch site to observe. Like many gallinaceous birds, the gavial hatchlings incorporate the remainder of the yolk into the alimentary canal for food during the first day or two of activity. Young gavials are cared for by both parents after they enter the water and for the next several months until they can take care of themselves.

The economy of metabolism dictates that birds be neither ovoviviparous or viviparous. Birds could not afford to expend energy in flight if embryos were developed in the oviduct. Therefore, birds are completely oviparous. In the interest of energy expenditure, eggs remain in the avian oviduct for the minimum time necessary for egg development. Not surprisingly, then, only one ovary and one oviduct become functionally developed in most birds. In some reptiles, however, where weight is not a factor, the eggs are not laid but are retained within the oviduct to be born at the proper stage of development. This ovoviviparous condition is common to many serpents.

There are marked differences, however, between birds and reptiles. Reptiles are poikilothermic (cold-blooded), and birds are homoiothermic (warm-blooded), although the warm-blooded characteristic develops only after some hours or days in the newly hatched young of birds. Only birds have feathers. Besides making flight possible, feathers aid in preserving heat in the body and are therefore necessary for a successful reproductive cycle.

Birds also show great advances over reptiles in broodiness, incubation, care of young, and nest building. These behaviors demonstrate a much higher level of development in birds over similar processes and functions in reptiles. The bird's warm-bloodedness, flight, and efficient mobility in the air and on the ground have naturally led, through evolutionary adaptation throughout the world, to complex life histories and a great diversity of bird life. On the other hand, many reptiles, cold-blooded and flightless, cannot adapt to environments in which many birds have adjusted successfully.

HOW REPRODUCTION BEGINS

Early manifestations of the start of the reproductive cycle are internal: hormones stimulate the enlargement of the gonads and result in the development of viable, fertile eggs and sperm.

At the same time, external changes take place in the bird that provide evidence of the early stages in the cycle. These are largely behavioral and include such processes as courtship, selection of territory, and nest building. (The nest must be finished and habitable when the egg is laid.)

Additionally, some species experience striking morphological changes. In many shore birds, for example, the dull coloration of winter feathers becomes bright, shining, colorful spring plumage. In gallinaceous birds spring feathers brighten and the comb and wattles become deep crimson. Bill colors may also be transformed, as in the myna and starling, from black to bright yellow. (Tennyson remarked that "in the spring a livelier iris changes on the burnished dove.") Darwin believed, as do many ornithologists today, that the brilliant colors of the male bird in the spring denoted inner health; therefore the female's selection of the more brightly colored, healthier male would tend to pass on to succeeding generations the genetic characteristics of the highest quality.

The age at which reproduction begins is extremely variable in birds. Many species start their first cycle at one year of age, while other species may not nest until they are eight to ten years old. A random number of

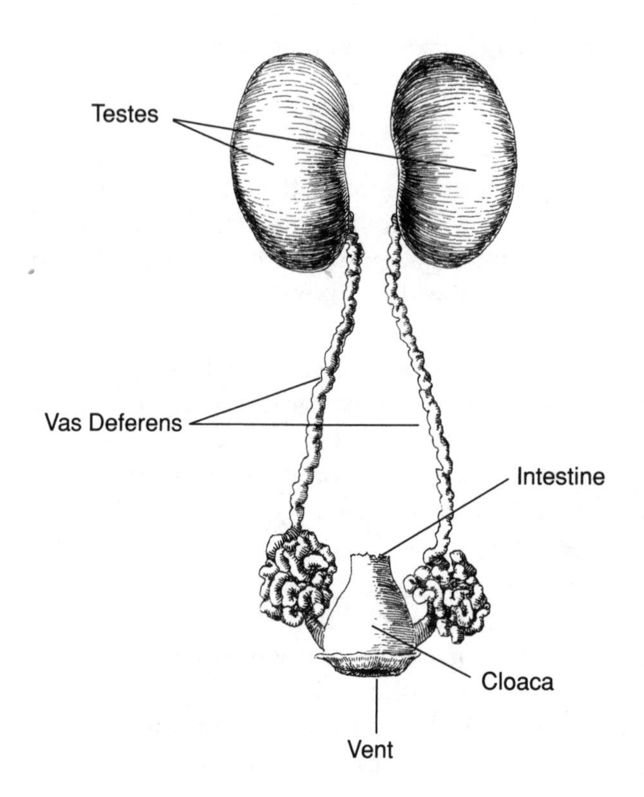

A male bird's reproductive system.

breeding terns collected over a four-year span indicated either a delayed first reproductive cycle or a long reproductive life for that species, because they ranged in age from four to fifteen years. Latitude and migration patterns modify cycles, as does maturation of physiological processes such as hormone development and growth of sex organs.

The testes in male birds are internal and forward of the kidneys. Each testis is oval or bean-shaped and subject to great variations in size and position depending on seasonal activities. During the reproductive cycle, the testes may migrate toward the posterior of the body and come to rest between the abdominal air sacs. During the sexual cycle the testes may enlarge as many as several hundred times. The testes of a duck during the breeding season can approximate one-tenth of its body weight.

The spermatozoa of birds, which exhibit a great variety of sizes and shapes among species, are very vulnerable to high temperatures or changes in temperature. The testes in birds are therefore cooled by an exchange of heat with air in the abdominal air sacs. The rapid and continuous exchange of air in the sacs can dissipate a great deal of heat through the lungs into the exhalent air. The temperature is thus kept a few degrees cooler around the testes than it is throughout the body cavity. Nevertheless, the attrition rate for sperm in birds is probably higher than in other warm-blooded vertebrates. To compensate, the number of sperm produced by birds must be correspondingly high. It is estimated that as many as 3.5 billion sperm are deposited in the cloaca at each copulation; the number may be as many as 8 billion in the domestic chicken.

The sperm develop in the testes and at maturity pass through a series of tubules and eventually into the coiled vas deferens leading to the cloaca. Between the testes and the vas deferens is a tubule called the epididymus, which probably aids in the maturation of the sperm. In birds this structure is small compared to those found in mammals, probably because birds store sperm in the vas deferens instead of the epididymus as mammals do. In some bird species the caudal end of the vas deferens enters into a cloacal swelling, which provides a cooler body area for the sperm.

Much is known about avian sperm because of research on the domestic chicken and other domestic species such as the turkey. Chickens lend themselves to successful research better than do most wild birds. In chickens, the sperm are developed and males are fertile at twenty-four to twenty-six weeks of age; the number of sperm in a single deposition varies from six to eight billion. The volume of each ejaculate in the chicken may approach a maximum of one cubic centimeter. For optimum fertility, a minimum of 100 million sperm must be deposited in the oviduct of the chicken; the sperm remain viable for as long as one month. The volume of ejaculate in turkeys is much lower, but the concentration of sperm is much higher in this species, so that the actual number of sperm per unit is higher.

At ovulation sperm must be present in the upper reaches (magnum) of the oviduct in order to fertilize the egg, because the egg does not move by itself. (If the egg is not fertilized it may be reabsorbed by the oviduct.) Sperm may be stored for a considerable time in the lower oviduct, but to reach the newly developed egg must travel the entire length of the oviduct. This period of travel may be as short as thirty minutes. The time between fertilization and deposition of the first egg in the nest varies widely in birds; it may be as short as eighteen hours or as long as several months, as in the brush turkey. In the chicken the average time elapsed between copulation and egg laying is between twenty and seventy-two hours.

Rarely does the adult female bird in any species possess two ovaries. The right ovary and oviduct are present in the embryonic stage but usually disappear before hatching. The loss of one half of this original organ system is probably of adaptive value; it reduces weight in the adult bird. In addition, if both ovaries were functional and each oviduct contained a mature egg, the parallel descent and pressure of the eggs on each other might distort their shape and ultimately destroy their shells due to the confines of the body cavity. In a very few birds—such as some ducks and the flightless kiwi and ostrich—two functional ovaries remain in the adult.

The ovary in the female, like the testes in the male, enlarges greatly during the breeding season and shrinks during the remainder of the year. In the non-breeding part of the year, such a tiny ovary is often difficult to observe in the dissected body. The pituitary gland in the brain develops a luteinizing hormone that stimulates the growth and enlargement of the ovary and controls the discharge of the ovum from the follicle when the ovum is ready to become fertilized. A single ovary, stimulated by another hormone from the pituitary gland, may produce as many as twenty-five thousand follicles, each of which is the beginning of an egg or ovum. But relatively few follicles ever develop into eggs; the remainder never become functional.

A bird's unfertilized egg, or ovum, is basically a large single cell consisting mostly of yolk. As the ovum bursts from the ruptured follicle, it is drawn into the infundibulum of the upper end of the oviduct. The point at which the mature egg develops may not be constant in all species. Nevertheless, the much-studied reproductive

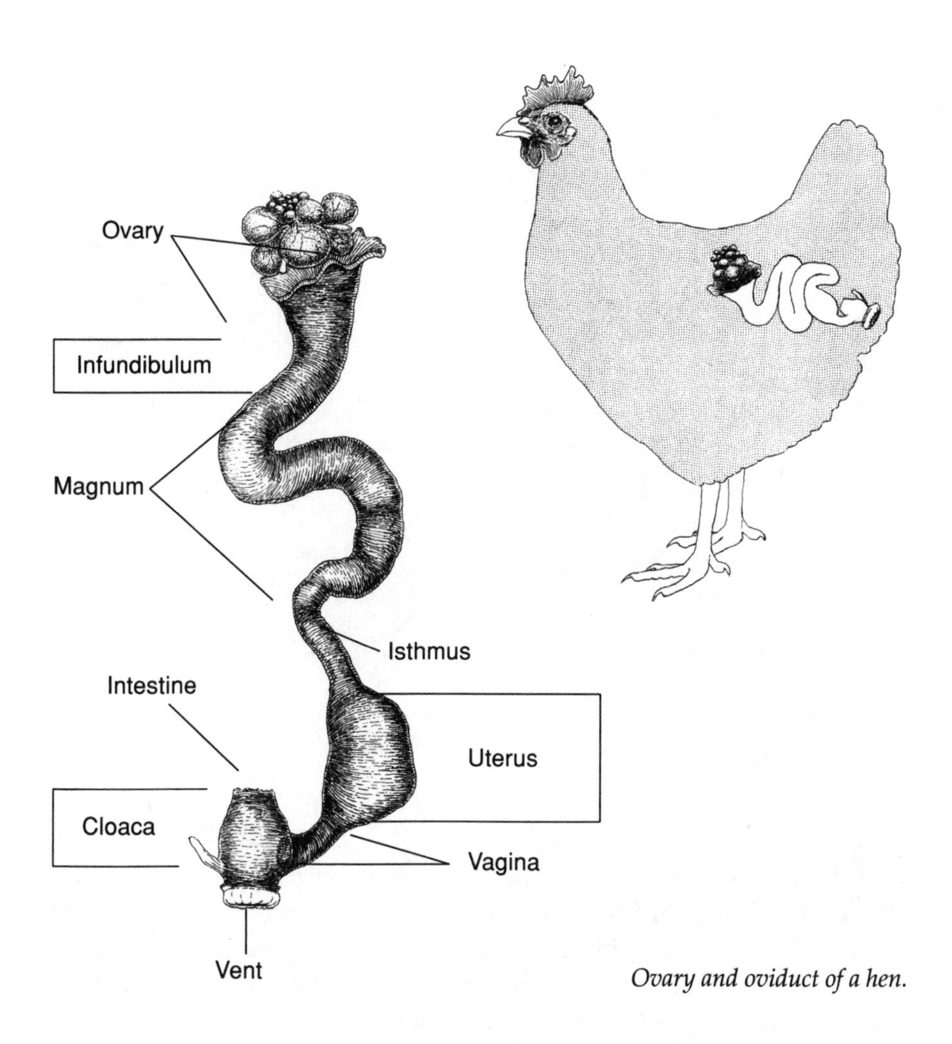

Ovary and oviduct of a hen.

cycle of the domestic chicken provides an example of a typical cycle. Within minutes after the egg is fertilized it begins its descent through the chicken's oviduct. Cell division begins immediately after fertilization and continues in the egg as long as it is kept warm.

The oviduct is a coiled, twisted tube with two muscular layers and an inner glandular layer. The muscles, by peristalsis, move the developing egg down the tube at the same time that the glandular layer secretes substances to add to the egg. Within the oviduct are

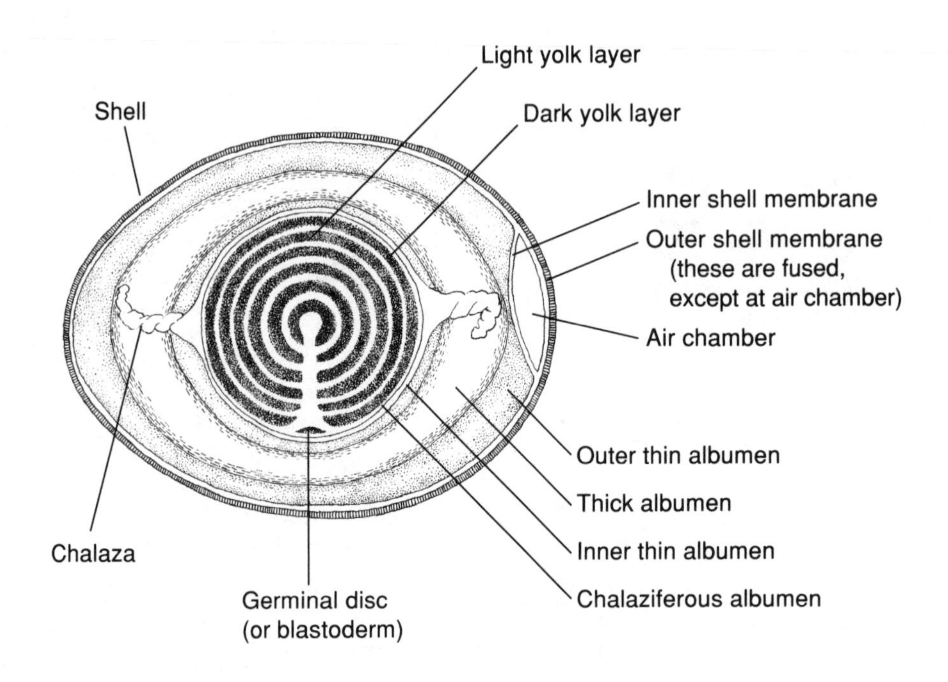

Cross-section of an egg.

five definite visible regions. Beginning at the ovary they include: the funnel-shaped infundibulum; the magnum, which is the largest region; the isthmus, which is actually the narrowest portion of the magnum; the large, muscular uterus; and the vagina, a narrow muscular section that aids in the laying process.

It may take several hours for the fertilized ovum (zygote) to pass slowly through the magnum, where it is surrounded by several layers of albumen, which will provide water and energy for the developing embryo. Inner and outer membranes envelop the descending egg as it passes through the short, narrow isthmus. The egg then enters the uterus where it acquires the external shell. In the chicken, the descent of the ovum from infundibulum to uterus takes more than four hours. The egg may be retained in the uterus for a period of twenty hours or more as the shell and pigments are developed. Uterine glands give the shell the pigments that are characteristic of the bird species. In certain species the eggshell receives only a surface film of pigment, while in others the shell color is homogeneous.

The entire structure is now a mature egg, and it passes quickly through the muscular cloaca to be laid. In the domestic chicken, it takes about twenty-four hours from the time the follicle ruptures until the egg is laid. If incubation begins and is later interrupted, the cooled egg can be warmed up again within a few days. As a general rule, though, if warming doesn't recommence within a very few days the cooling is fatal.

The white of the egg is largely composed of proteins; it is about ninety percent albumen and ten percent traces of other proteins. Several thin, watery layers of protein alternate with denser, sticky layers. The egg rotates as it descends through the oviduct so that the central thick albumen layer surrounding the yolk becomes a twisted structure called the *chalaza.* This keeps the yolk suspended in the white. The yolk may then rotate freely in such a way that the germinal spot formed by union of egg and sperm cells is always oriented on the dorsal surface of the yolk. The white of the egg is used largely for tissue growth in the developing organs of the embryo. Fats are contained mostly in the yolk and are used primarily for energy. The outer shell of calcium carbonate is porous, permitting the exchange of gases such as carbon dioxide and oxygen. Water vapor also passes through the shell to the outside. The phosphorous and sulphur that are present in the egg yolk are inorganic constituents that help form the organs in the developing bird. The percentage of yolk varies greatly in different species, but altricial birds tend to have small percentages of yolk while precocial species have larger percentages, since precocial birds are in the egg longer and are more developed at hatching.

While the eggs are developing, selection of the nest site can be proceeding.

THE COURTSHIP

The courtship phase in the life cycle of birds does not follow a definite pattern that is the same in all species. In birds that are migrating from the winter grounds to the nesting grounds, they actually start their courtship prior to the migration. This is probably a benefit to the species because they are able to begin their nest building upon arrival in the nesting area. In species that must migrate to a nesting ground, selection of territory would obviously occur after courtship has started.

Song

Birds communicate with each other by various means, including visual recognition and sound. One of the main ways, of course, is through song. The word "song" implies something that is pleasing to the human ear, like the twitters and chirps of the passerines, which many ornithologists and bird enthusiasts have carelessly fallen into the habit of calling "songbirds." But really the strident cries of the crow or raven, the "hoot" of the owl, the "quack" of the duck, the "honk" of the goose, and the "gobble" of the turkey—or any sound produced by the syrinx in birds—also ought to be classified as song. When they are uttered in their normal habitats, they take on great significance to a bird species, especially during the breeding season.

In general, male birds tend to sing more than females and tend to have more complex songs. Sounds from the respiratory system in birds originate in the syrinx just forward of the lungs. The trachea, or windpipe, enters the syrinx at the base of the neck, and two bronchial tubes extend from this structure, one to each lung. The syrinx is constructed of light, porous bone with "windows" of thin, resilient membranes. The size, shape, and number of windows is characteristic of a species. Air forcibly ejected from the lungs sets up vibrations in the window membranes. Many songs are modified by movements in the throat and bill. Pitch is regulated by the number of vibrations set up in the window membrane, the overall size of the syrinx, and the volume of air in the syrinx. The size of the syrinx is not related to the size of the bird. The syrinx in a merganser duck, for example, is many times larger than the syrinx in the much larger eagle. This explains the difference in pitch between the shrill scream of the eagle and the low-pitched song of the duck.

Most birds make ample use of songs and song-producing mechanisms during the courtship phase of the reproductive cycle. Early in the breeding cycle the male bird usually selects a territory and a strategic spot within it to proclaim dominance within its territorial limits. Continual singing by the male establishes this territory and entices the female to the area.

Many ornithologists contend that birds recognize the songs of their species and subspecies. For identification purposes, some students of birds now suggest that close attention should be paid to the song characteristics of a subspecies rather than just the morphological or physiological characteristics. Two subspecies might not interbreed because each does not recognize the voice of its sympatric neighbor. Thus, voice may be an important factor in the processes of courtship, and it may aid in reproductive isolation. Consequently, voice becomes part of the mechanism that helps stabilize characteristics in a species or subspecies.

Structures other than the syrinx may be used to develop sound and may also serve as showpieces during the courtship phase. A number of gallinaceous species possess inflatable air sacs on the sides of the neck immediately forward of the shoulder. These are used to produce sound during complex, ritualistic dance displays. In the blue grouse (*Dendragapus obscurus*), these air sacs produce a booming sound. The male blue grouse continues this activity into the

A male frigate bird inflates his red gular pouch to get the attention of a female.

and is used by the male to attract the female during courtship. No sounds emanate from this pouch, but it is apparently important in courtship as well as in demonstrating dominance in a group.

While display activities are most often indulged in by the male, they are not unique to him. The female may exhibit her own distinct displays—alone or in concert with the male. The female phalarope, for example, performs most of the courtship displays: she flutters over the male as he sits on the ground.

With western grebes (*Aechmophorus occidentalis*) and some waterfowl, male and female perform their dances together. Pairs of grebes arch their necks together and literally walk across water with their strong "paddlewheel" feet. The awkward pose and flailing movements are showy, if not beautiful. Some waterfowl, with rapid foot movements, tread water with arched neck and one wing held high. Elaborate steps and movements can reach a high level of complexity, as with the birds of paradise, some species of which hang head down from the limbs of trees while folding and moving their various feathers and singing.

Most dance activities are vigorous and animated and can extend over a considerable period of time. In most species of birds in which the dance is elaborate, the male is the principal or sole participant, though his dance activities are stimulated by the presence of the female. Where sexual dimorphism is the rule, the bright colors of the male and the more modest colors of the female may make it easier for one sex to recognize the other, but may also stimulate the male to a rising fervor of activity because he easily recognizes the female.

The ultimate colorful and complex display is found in courtship rituals performed in an arena—usually called the lek. A lek is an area in which two or more male birds stage their display dances, accompanied by various sounds that are produced by vocal structures, vibrations, or movements of stiffened feathers. The display area may be small—a few square feet—and occupied by a few pairs. Or it can be large—an acre or more—and used by many males in concert.

In the lek the males are usually polygynous and the females promiscuous. Sage grouse (*Centrocercus urophasianus*) males set up small areas of territory within the lek area. The older dominant males choose the center, which automatically relegates the young males to the lek's periphery. The females usually accept only the males in the center. The result is that about ten percent of the males in the population account for seventy-five percent of the matings. A small percentage of males, therefore, is responsible for perpetuating the species. The young males occupying the periphery wait until they are allowed by age to move to the center. As Wilson (1975) says, "There is a waiting list for prime real estate."

Lek behavior is usually found in species that exhibit extreme sexual dimorphism. Some Central and South American birds and even a few North American birds, most notably the sage grouse, use well-developed leks. Not surprisingly, leks are found in a few species of bower birds and in two species of birds of paradise—usually regarded as the most beautiful of all birds.

The bower birds of eastern coastal Australia and adjacent New Guinea have developed especially elaborate nesting sites and courtship areas. The nest proper is usually a shallow saucer constructed of twigs and eucalyptus leaves and other vegetation. The female alone usually constructs the nest while the male is busy building an elaborate "courtyard," or bower, for display and courtship. This bower is usually not far from the nest. Most of the eighteen species build a bower of some kind, with walls of interwoven twigs, moss, and other materials. The bowers may be built tentlike around the base of a small tree or as tunnel structures several feet long.

The floor of the bower will contain a great variety of foreign substances, or perhaps a cache of materials and organisms, such as snails, from the immediate vicinity. These accessory artifacts of the bower bird represent one of the most elaborate ranges of substances used for courtship and nesting activities by any bird. Apparently such collections are necessary for the successful reproductive cycle to satisfy the instincts of the bird. Several species of bower birds are very fond of collecting and arranging snail shells on the floorway. Collected materials placed in the bower of other species may include pieces of glass, string, spent brass cartridges, colored plastic tubing, and toothbrush handles. One species builds its display with a roof of interlocking twigs as high as three feet and up to six feet in diameter. This species is careful to keep the floor of its domain carpeted by moss and completely clean of any debris, but native flowers and fruits are carefully placed at the entrance. The satin bower bird is highly susceptible to the color blue. Cooper and Forshaw, describing a bower of this species, noted that it was decorated with seventy-five blue plastic items, including tops from ball-point pens, plastic pegs, bottle tops, a toothbrush, and parts of broken toys, plus fifteen blue-edged feathers, ten dried snail shells, several pieces of fawn-colored garden twine, and two blue marbles.

Several bower birds "paint" the avenue of the bower with macerated leaves mixed with saliva and applied in lumps. It has been suggested that "painting" may provide a clue concerning ownership of the bower. Another species of bower bird develops its bower with a central "avenue" leading through the structure. The floor here is strewn with pebbles, white bones, and fruit. In all species, the male, while building and later performing in the bower, gives voice to various songs, cries, chirps, and "conversational" noises.

It is difficult to conclusively explain the reasons for the elaborate and detailed construction of the bower in this group of birds. While this behavior is certainly related to the courtship and display segment of the reproductive cycle, it is difficult to explain the necessity of this complex phase for the adaptive success in the whole range of the bower bird group. Perhaps the ability to build a better bower reflects a genetic superiority on the part of the male, who then becomes more likely to be selected by the female.

Pair Bonds

A bird's response to courtship behavior depends upon whether the species is monogamous, polygamous, or promiscuous. Monogamous species mate with one partner for one season or more, and some mate for life. They develop strong pair bonds. About ninety percent of all bird species are regarded as primarily monogamous. Because birds are warm-blooded and oviparous, they require extensive care of eggs and young. Consequently, it is more practical for two parents to work together to establish and maintain the reproductive cycle from nest building through fledging. The demands are much too great for a single parent, especially when a clutch numbers three eggs or more. In species where one parent can successfully rear the young, polygamy or promiscuity is the rule. Most vertebrates other than birds are either promiscuous or polygamous.

There are two kinds of polygamy: polygyny, where one male mates with several females, and polyandry, where one female mates with several males. Polygyny is the pattern in many gallinaceous species; polyandry, while not as common as polygyny, occurs in certain species that demonstrate a reversal of the normal sex roles. In the polyandrous phalaropes, for example, the female has the brighter plumage and is aggressive in its courtship, but does no nest building, incubating, or feeding. Such behavior is probably an advantage to this species, which nests in the northern tundra where the breeding season is short. The female lays her eggs in a nest,

leaves the incubation and care of young to the male, mates with another male, deposits a second set of eggs in another nest, and is thus responsible for a much larger addition to the phalarope population. Polyandry also may occur in populations of birds with an excess of males. In addition to the phalarope, polyandry is found in tinamous, jacanas, and some quail.

Some species of birds may shift back and forth from monogamy to polygamy. Although monogamy is the rule among most waterfowl, some species are polygamous, simply because the male is not needed to bring food to the precocial young that are led about by the female.

Promiscuous birds separate immediately after mating. In many species it is the male that deserts the female, leaving all duties in the reproductive cycle to her. The pair bond exists only long enough to insure fertilization. This type of behavior is conspicuously noted in species where the male displays brightly colored feathers of different sizes and patterns that contrast sharply with the relatively drab dress of the female. The departure, for example, of the brightly colored male hummingbird during incubation or feeding tends to reduce predation. In addition to hummingbirds, some grouse, pheasants, and birds of paradise are also promiscuous.

Mating

Mating usually occurs within nesting territories. But in migratory species, pair bonding is often established on wintering grounds.

The process of mating in most birds involves a brief touching of the cloacas of the mating pair. This has been called for many years the "cloacal kiss." The process is brought about by the male standing on the back of the female as the male twists his tail under the tail of the female. The entire process is completed in a few seconds. Sperm deposited in the female cloaca swim directly up the oviduct to the infundibulum. In domestic fowls this occurs in approximately thirty minutes. After a successful copulation, domestic turkeys and chickens can produce fertile eggs for as long as six weeks. This is remarkable since female birds have no sperm storage areas.

A few birds have an erectile intromittent organ. This special structure is found in tinamous, most waterfowl, storks, and ostriches. It is not clear why these few birds possess such an organ, but in waterfowl it may make copulation easier and more effective for sperm deposition underwater.

Copulation may occur on the ground, in trees, in the water, or on wires strung between fence posts or power poles. In a few species such as swifts and some swallows copulation occurs in flight.

TERRITORIES AND TERRITORIAL FUNCTIONS

A territory may be large or small. It may be the many square miles demanded by the needs of the golden eagle or it may be the small patch of blooming flowers selected by the hummingbird. Proclamation of the territory consists of songs and displays.

Birds tend to be more selective of territory during the breeding season than at any other time. Aggression increases in intensity before and during the reproductive cycle. Territorial selection for food may be observed in the winter season, at which time various species live together compatibly, but during the reproductive cycle most birds show a marked increase in aggressive behavior and pugnacity toward other species and each other. Fighting by males, whether of differing species or the same species, can be intense; sometimes such a duel is a fight to the death.

Staking out and protecting a chosen territory has distinct advantages. Defending a territory during the reproductive cycle serves various purposes: it protects the pair, nest, and eventually the young, guards a food supply, or simply protects the region from interference with the normal functions of daily living. It may also prevent too great an increase of the species in one area. Additionally, the territory usually contains the desired nesting materials and food suitable for the adults and young. This last advantage is speculative, however, since many birds gather food from far outside their selected territory. Although a robin in a backyard has everything it needs for its young close by, pelicans

The pipit (Anthus) *nest is large and bulky in order to retain heat in the subalpine territory where the bird nests. Life size.*

nesting on an island in the Great Salt Lake will fly a round trip of fifty miles daily to bring food for theirs. Some raptors, in order to find enough food for their broods, must range far and wide; one eagle's territory may cover one square mile or they may have to range several miles in order to find enough food to sustain them.

In some species, selection of territory appears to be related to time of migration. For example, the first migrant flocks of the black turnstone (*Arenaria melanocephala*) tend to be predominantly male. After some days have passed, the flocks are made up mostly of females. The male's early arrival on the nesting ground probably allows him enough time to select and defend a territory several

*The white-tailed ptarmigan (*Lagopus leucurus*) is a montane species and nests in the heather fields above timberline.*

days before the female gets there.

Although territories are usually selected by males, in some species the choice is made by the female. As mentioned before, the female phalarope exhibits this type of role reversal. The female of the rufous hummingbird (*Selasphorus rufus*) also selects the habitat and the territory; when the eggs have been fertilized the male leaves the vicinity and doesn't return until a second clutch is to be started. Each member of a pair of Lucifer hummingbirds (*Calothorax lucifer*) selects a separate territory and vigorously defends it until the nest is about to be built on the area selected by the female. For some unknown reason, the male may continue to defend his own selection for a short time but will soon abandon it.

Various species of seabirds acquire and defend territories with similar characteristics. For example, puffins, petrels, and auklets all nest in burrows in banks adjacent to salt water. For these birds, selection of territory is not entirely based upon need for available building material, nor upon a need for protection from predators, but rather for the convenience of an adjacent food supply. Where nesting and feeding areas are sufficient, competition is minimal, usually because feeding methods differ. Puffins and auklets feed on small fish that they secure by diving; petrels, albatrosses, and shearwaters are surface feeders. There is some competition between various species of terns and gulls, but again their different feeding habits prevent the competition from becoming too stiff.

Most seabirds nest in colonies, but a few gull species will not live compatibly together. The glaucous-winged gull and the western gull may nest on the same island, but the nest sites are far away from each other because the larger gulls may attempt to feed on the young gulls and flightless terns. The California gull and ring-billed gull, however, will nest side by side in a Caspian tern colony. The terns, because of their food habits, will not prey on the young gulls. If there is little or no competition in acquiring food, colonial nesting is very advantageous since it provides safety in numbers.

The Clark's nutcracker (Nucifraga columbiana) *chooses a territory in the alpine regions early in the season while there is still much snow. The massive nest helps retain warmth for the incubating eggs. One-half life size.*

SELECTING A NEST SITE

Birds, through the centuries, have become adept at successfully selecting a great variety of habitats and locations within their territory for making nests and laying eggs. Underground, at sea level, at the snowline in the mountains—various species occupy these and all the habitats and niches in between. Selection of the proper habitat within the territory is critical for the success of the breeding season. Habitats must provide protection from predators and a secure spot to hide the nest; for instance, many nests are camouflaged or hidden in inaccessible places. Eggs or nests placed high in trees, in burrows, on cliff ledges or islands are difficult to locate and approach. Habitats must also provide materials for nest construction and in some measure make a nest less subject to destruction by wind, weather, and other hazards.

Birds are remarkably consistent in selecting a particular site. Colonial nesting seabirds return to the same area year after year; other species return to their selected habitats to build new nests on the remnants of last year's nests. But there is little consistency in site choice among members of any one family. For instance, in the swallow family (Hirundinidae), tree swallows (*Tachycineta bicolor*) and violet-green swallows (*Tachycineta thalassina*) seek out convenient nest cavities in trees; cliff swallows (*Hirundo pyrrhonota*) place their gourd-shaped mud nests in great colonies on cliff sides, bridge girders, or walls of buildings; barn swallows (*Hirundo rustica*) use mud to construct nests on girders,

A coot sits quietly in its ground nest. As long as it does not move, it is concealed and fairly safe.

rafters, or cornices of buildings; bank swallows (*Riparia riparia*) burrow into banks adjacent to water.

No one order of birds has any monopoly on the kind of site selected. Some seabirds and kingfishers, like the bank swallows, also dig burrows in banks. Many passerines nest in tall trees, as do some herons belonging to a strikingly different order. Most shore birds nest on the ground, but so do many passerine birds. The Piciformes nest in holes or cavities in trees, as do certain ducks and some passerines.

Ground Nesters

There are many birds that place their eggs on bare ground or in the open without the benefit of a nest structure. This may be explained in part by the fact that these birds do not have the physical capabilities or available materials to build nests. Among the

*The glaucous-winged gull (*Larus glaucescens*) nests at sea level and forms a large bulky nest of grasses and dried seaweed. One-half life size.*

Procellariiformes (albatrosses and petrels, among others), the nest is a mere scrape in the sand, as are the nests of some species of terns and tropic birds. Some species of penguins hold their egg on top of the foot because the ground is covered with ice in the antarctic. Murres lay their eggs on rock ledges adjacent to the sea. (Fortunately, the eggs of these species are pear shaped; they roll in a tight circle and do not fall into the sea.)

Birds such as grebes, tube-nosed seabirds, most waterfowl, gallinaceous birds, shore birds, coots, cranes, and rails consistently build their nests or place their eggs on the surface of the ground. This is also the favorite habitat of many passerine birds. Ground nesters are found in a great variety of habitats out in the open.

Gulls, terns, and some shore birds use the sand or pebbles on the beach, scraping out a shallow cup and placing eggs on bits of shell, seaweed, and grasses. The open tundra of the northland is the nesting ground of many other shore birds that use the cup nest as well. Colonial nesting seabirds commonly use islands on which to lay their eggs. The selection of such a site results in a degree of protection. Beck, collecting sooty tern (*Sterna fuscata*) eggs on Clipperton Island, Mexico, on August 10, 1905, recorded that the two eggs he took on this date were dropped with one thousand others on a coral islet that measured only twenty feet by forty feet (UPS Museum). Besides the protection offered by the isolation of the island, the extreme concentration of nesting adults served to further safeguard the population, as there is protection in numbers.

Cormorants build on ledges high on cliff sides next to water. (Some cormorants may build their nests on dune islands a few feet above sea level or on a navigational beacon.) A few raptors, such as the ferruginous hawk (*Buteo regalis*) and the prairie falcon (*Falco mexicanus*), seek out rock ledges overlooking wide valleys, while the peregrine falcon (*Falco peregrinus*) nests on cliff sides adjacent to the sea. A few dove species build shallow nests in open ground. Some owls, including the snowy owl (*Nyctea scandiaca*), have developed the ground-nesting habit out of necessity because there is an absence of suitable trees in their tundra home. On the shrubby ground in the Upper Sonoran desert can be found some species of sparrows and the long-billed curlew (*Numenius americanus*). Savannah sparrows of the open fields hide their nests among the grasses of dunes and adjacent to brush borders.

The open prairie is a favorite location for horned larks, meadow larks, vesper sparrows, and savannah sparrows. The killdeer (*Charadrius vociferus*) builds its nest on open gravel or on the slope above a favorite lake shore. The pipit (*Anthus spinoletta*) seeks out heather meadows at considerable elevations in the mountains. The rosy finch (*Leucosticte arctoa*) places its nest on a rock ledge overlooking alpine flower fields high in the foothills and mountains where it becomes neighbor to the pipit and ptarmigan. The circumpolar Lapland larkspur (*Calcarius lapponicus*) and the snow bunting (*Plectrophenax nivalis*) nest on the ground from sea level to higher

*The rosy finch (*Leucosticte arctoa*) is a subalpine nester with a large bulky nest for heat retention. One-half life size.*

*The yellow-throated warbler (*Dendroica dominica*) places its nest in a large clump of Spanish moss, which it uses to construct the structure. The bird doesn't have to travel to find nest materials, and the nest is well camouflaged. One-half life size.*

*The Virginia rail's nest (*Rallus limicola*) is an "island" in the shallow water of a suitable swamp. One-half life size.*

elevations in the far northern mountains north of the treeline. The bunting outdistances the larkspur in latitude; it nests the farthest north of any land bird in the world.

The forest floor is a favorite habitat for nest building for many birds, including the quail, grouse, ptarmigan (*Lagopus leucurus*), and lyrebird.

Several warbler species bury their nests in soft moss beneath trees or in low bushes a few feet from the ground, as do sparrows, juncos, and towhees.

The nests of some ducks can be found on the ground adjacent to the swamp or shoreline. In the rush clumps lives the yellow-throat (*Geothlypis* sp.); in a nearby wet area, on a six-inch platform of vegetation with a concave cavity at its top, sit the eggs of the Virginia rail (*Rallus limicola*). Coots, rails, and grebes build their nest structures of reeds and grasses on the shores of lakes or swamps. Some of their nests even float on open water. The industrious coot (*Fulica americana*) very carefully ties its floating "raft" to an adjacent shrub.

Many species burrow underground for nest shelter; these are usually birds associated with the sea. Subterranean nesting eliminates the need for an elaborate nest; most eggs are laid on the bare floor with little or no nest material. Such seabirds as auklets, puffins, shearwaters, petrels, and some penguins go to great lengths to dig a hole or elaborate labyrinth of tunnels in the bank adjacent to an ocean. Some of these burrow-nesting birds show remarkable adaptations to this underground type of nesting. Auklets, for example, develop long, sharp claws that help them dig tunnels.

The Atlantic puffin (*Fratercula arctica*) uses a ready-made burrow for its egg. The burrows of resident rabbits, which honeycomb the arctic slopes, are regularly taken over by nesting puffins in great colonies.

Storm petrels (*Oceanodroma* sp.) of the northern hemisphere build their complex labyrinthine tunnels in layers at the root levels of hardy shrubs. It is often necessary for these species to dig around the tough roots of the plants. These plants, worn by the prevailing winds, develop astonishing mats of thick "crew-cut" limbs that are virtually impenetrable. At night when the adult petrels come forth to feed, their flutelike voices can be heard more and more clearly as the birds struggle upward through the tough tangle of vegetation. Even on a dark and moonless night, the parent bird makes its way back to the egg by penetrating the thicket—perhaps a couple of feet deep—to the correct tunnel opening at the base of the plant stems. The air above the hidden colony can be alive with hundreds or even thousands of adult birds, some just emerged from the nursery and some finding their way back to the proper entrance. Inside the winding tunnels single eggs are laid in tandem along the floor, each constituting a clutch. Since the labyrinth is occupied by a number of adults, the eggs are spaced at some distance from each other in a sort of underground territory system.

While most penguins in the southern range of the antarctic place their eggs on the surface of the ground, colonies of Australia's fairy penguins (*Eudyptula minor*) construct burrows in the sand among the roots of dune grasses. The kingfisher, which also puts its clutch in a tunnel in a sandbank, may be an immediate neighbor to a guillemot, which lays its eggs in sand burrows or right on the sand at high water amidst the "jack-straw" tangle of large drift logs. Although these two species, like many birds, are highly territorial, especially in the breeding season, they do not compete with each other; the guillemot dives in deep water for its fish and the kingfisher hunts along the shallow water at the edge of the beach.

Of those birds that place their eggs on bare ground, some use considerable material and develop elaborate nests while others use little or no material. In gallinaceous birds, the nest material is abundant but the nest is loosely formed. It may consist of grasses, leaves, or feathers placed on the forest floor, or it may simply be loosely constructed of leaves, other vegetation, and soft underfeathers of the adults.

Ground-level nests with eggs are more vulnerable to predators than eggs placed high in trees or in nest cavities. Gallinaceous birds and waterfowl tend to "sit tight" and depend on their coloration to conceal and protect them. When leaving the nest quietly for routine feeding, the adult waterfowl carefully hides the down in the nest by covering it with leaves.

Although modern birds exhibit a great variety of habitats for the placement of the eggs—with or without a nest—there is one environment that will not support incubation of the egg. No modern bird lays its eggs directly into water. When eggs laid near the water's edge become covered with water, the embryos are destroyed. Under the pressure of population increase in a colony of seagulls, for example, eggs are commonly laid at the periphery of the colony on an intertidal beach. The first tide that covers these eggs promptly destroys them. This can also happen to eggs laid on the banks of rivers or estuaries when the water rises due to unusual snow melt or heavy rainfall. Nevertheless there is evidence that the fossil *Hesperornis* of Cretaceous seas gave birth to its young at sea. This ancient bird could not fly and was very clumsy on land, but it could swim great distances between land masses. It is probable that *Hesperornis* was ovoviviparous. The egg that hatched in the oviduct

*All of the Piciformes, such as this yellow-bellied sapsucker (*Sphyrapicus varius*), use a liberal quantity of chips in the nest cavity. Life size.*

was protected from direct contact with the water during incubation. In a modern egg submerged in water, incubation temperature cannot be maintained, and respiratory processes fail.

Tree Nesters

Piciforms select a wooded territory dotted with dead trees that can be used for cavity construction. Nesting material is not necessary within the cavity; the eggs are placed directly on wood chips on the chamber's floor. Thus, selection of such a habitat is not made on the basis of availability of twigs, leaves, mosses, or other nesting substances. The major requirement here is the presence of dead trees. A comparable situation exists in the southwestern deserts of

The eggs of the hairy woodpecker (Dicoides villosus) *resting on a bed of fine wood chips. Life size.*

America where the elf owl (*Micranthe whitneyi*) is quick to use the abandoned nest cavity made by a desert woodpecker in a live saguaro cactus. No nest material is needed. When the woodpecker invades the cactus, the sap of the living plant develops a hardened lining, forming a perfect egg cavity for the tiny owl a year later.

Other birds that nest in cavities do require material to construct a nest within. Examples include the nuthatches and the chickadees. These birds hollow out a cavity inside a dead tree and construct a small entrance hole, then build cuplike nests of moss and other forest vegetation on the floor of the cavity. There is one notable difference between nuthatches and chickadees: the nuthatch's territory must include living resinous trees as well as some dead trees. Pitch from the living tree is gathered by adult birds and applied to the

A pair of northern flickers at work on their cavity nest.

outside of the entrance hole. This natural "flypaper" acts as a barrier to crawling insects that might disturb the nesting cycle of the rightful owner. Nuthatches gather pitch throughout incubation, and consequently a considerable amount of pitch accumulates around the nest entrance toward the end of this period. When entering the nest cavity, the nuthatch proceeds into its entrance hole upside down; this way it picks up less pitch on its feet, because the substance tends to run down the tree; little remains plastered around the upper part of the hole.

Swifts also use the inside of old trees for roosting and nest sites. Swifts are so unusual among birds that they warrant a little special attention. Appropriately, they are among the swiftest flyers, being capable of speeds of eighty miles per hour while feeding and perhaps two hundred miles per hour during a dive. The swift probably spends more of its life in the air than does any other species of land bird: it is in the air that they court, gather insects for their young, drink and bathe (in the rain), copulate, and collect nest twigs. The anatomy of the swifts is highly modified for the reproductive cycle. To accommodate their nesting sites inside old, abandoned chimneys or hollow trees, they have developed short, stubby legs with sharp claws and small spines at the end of the tail feathers, which help the swifts climb up the inside of a tree, much as woodpeckers or creepers scale a tree's exterior. (Similarities in beaks, feet, and spiny tail feathers represent a classic example of convergent evolution in the widely separated orders to which swifts, woodpeckers, and creepers belong.) When ready to leave the nest, young swifts crawl up the inside of the stack or tree, perch for a short time, and then take off on their own.

In the swifts, the salivary glands enlarge during the breeding season to secrete a thick, sticky saliva. The swifts of North America gather small, matchstick-sized twigs that they break off with their feet while in flight. Using their sticky saliva, they construct a shallow and shelflike nest inside a tree or chimney. W. E. Unglish collected and described a nest and clutch of three white-throated swifts (*Aeronautes saxatalis*) on May 18, 1936, in Santa Clara County, California (UPS Museum). This set, placed in an old swallow nesting hole, was a conglomerate of sand, grass blades, and owl and chicken feathers—all cemented together. Some of the swiftlets of Asia actually construct a nest on the underside of leaves of certain shrubs and trees. The nest is built in the form of a pad of feathers glued together with salivary secretions. The eggs are in turn glued tightly to the nest pad in a vertical position. To incubate the eggs, the parent birds cling to the nest pad with their claws and sit upright. Even when the wind moves the leaves on which the nest is constructed, the eggs stay firmly attached. The swifts of Asia don't use twigs to form their nests but instead build them entirely of translucent, salivary secretions. This nest, when dried, becomes an article of commerce which is sold in markets and restaurants to make bird's nest soup, regarded as a delicacy by gourmets in the parts of the world where these swifts are found.

*Nest and eggs of the chimney swift (*Chaetura pelagica*). The saliva that glues the twigs together can be seen in the lower right corner of the nest. Life size.*

Some old-world swifts, instead of attaching their nests to the side of a chimney or hollow tree, glue their nest material together to fashion a crude nest on the floor of old, abandoned buildings. The black swift (*Cypseloides niger*) of the Pacific coast of North America attaches its nest to a rocky wall, sometimes using a little mud in addition to saliva. The small-crested swift (*Hemiprocne longipennis*) of southeastern Asia glues a tiny nest of feathers on the top of a limb high in the tree. A single egg is glued to the center of this miniature nest.

Many species of birds from several orders consistently use the understory of shrubs and the lower branches of trees as nest sites. Bushtits, robins, orioles, waxwings, kinglets, warblers, vireos, some sparrows, and many other passerines typically choose shrubs and

The goldfinch (Carduelis) *prefers to place its nest in the crotch of a tree. Life size.*

lower limbs of trees. Their nests can be fastened to the limb of a tree next to the trunk, saddled out on the branch away from the main stem, built in the crotch of several branches, or tucked away under drooping branches and tied to the tree with cobwebs.

The nests of some species may be characteristically placed in quite exact places in vegetation (though many species will utilize different locations as substitutes if their favorite site is not available). For instance, the nest of the yellow warbler will usually be found in the crotch formed by several branches in a shrubby area. The bushtit weaves the upper portion of its pensile nest around the end of a limb in a shrub or evergreen tree. Waxwings, thrushes, and robins build their nests saddled on a horizontal limb toward the main trunk of the bush, as do certain sparrows and juncos. The night

*The gnatcatcher (*Polioptila caerula*) places its nest in the fork of a tree and covers it liberally with adjacent lichens. Life size.*

heron constructs its home on the ground in marsh vegetation but will on occasion build its nest in a shrub a few feet off the ground. The winter wren is very versatile and will use a convenient shrub for a nest but prefers to build among the tangled root mass of an overturned tree deep in the forest. This wren, almost alone among birds, chooses a site in the darkened interior of the dense forest. Blackbirds prefer the security of a shrub growing out of water. The marsh wren also uses the water habitat, usually building in cattails or other herbaceous plants instead of shrubs. For protection from predators, such species as the curve-billed thrasher (*Toxostoma curvirostre*) and the northern mockingbird (*Mimus polyglottos*) seek out trees or shrubs heavily armed with sharp spines or thorns.

*This nest of the curve-billed thrasher (*Toxostoma curvirostre*) was placed in a cholla cactus. The thorny plant serves as protection for the nest, eggs, and eventually the chicks. One-half life size.*

Many birds construct their nests in trees. Herons nest in colonies, often high in the trees of a forest, which is a nidological peculiarity. The adult heron flying into a nesting region carrying sticks for the nest or fish for young birds is an incongruous sight: the bird's long, wading legs and long, curved neck seem out of place amidst the tops of trees. Their nests may be as high as two hundred feet up in large, old-growth trees. The osprey finds an old tree with few if any limbs remaining and places its nest at the very top. Its nest is a bulky structure made of branches and perhaps lined with some leaves.

Some birds build new nests on the remains of last year's nest. Some owls, for instance, will use the abandoned nests of other species (with a little repair work). Old hawk nests are favorites, if available, and some owls, such as the screech owl and saw-whet owl, are willing to utilize natural protective cavities in trees. Hummingbirds also use former nests as foundations. Three- or four-decker nests are not uncommon. The hummingbird's nest may be built in shrubs, but large trees with pendant limbs over a clearing are another favorite spot for these colorful birds.

Some small comment should be made here to point out the great variety of nesting habitats of owls (Strigiformes). While most orders exhibit a fair degree of consistency in habitat types, owls show extreme variation in nest site selection.

The snowy owl in the far North nests on the ground in tundra. South of the snowy owl habitat, the great horned owl nests in trees. In temperate climates across America, screech owls, saw-whet owls, long-eared owls, and others nest in shrubs and trees some distance off the ground. Short-eared owls are quick to nest on the ground in these temperate areas.

In drier regions across the country, the burrowing owl seeks out old gopher holes and places its eggs underground. Still farther south, the diminutive elf owl almost exclusively utilizes old woodpecker holes made in saguaro cacti in Southwest deserts.

Of much concern today is the destruction of the habitat of the spotted owl. Its normal nesting region is deep in the old-growth forests of western America. One of its favorite environments is the temperate rain forest of the northwest corner of Washington state, which may have as many as 150 inches of rain or more per year. As the forest products industry harvests the old-growth trees, spotted owl populations decline.

The barn owl, not defeated by human civilization, is quick to take advantage of a man-made structure such as a barn or belfry, which houses abundant food, such as rats and mice, for a hungry owl brood. This owl, probably one of the world's most widespread species of birds, is still extending its range to the north, following the spread of human populations and man-made structures.

Other Habitats

There are other birds that find no difficulty nesting successfully in areas dominated by the noise and movement of human activities.

For example, several pairs of peregrine falcons have nested successfully on bridge girders in New York harbor and a few falcons have taken up residency in busy downtown Los Angeles. Violet-green swallows, house sparrows (*Passer domesticus*), and house wrens (*Troglodytes aedon*) seek out convenient cavities under the eaves of a dwelling. Martins will establish a colony on the side cornices of a tall downtown building.

Some non-migratory races of the Canada goose (*Branta canadensis*) seek out open places adjacent to man-made recreational facilities. These regions are usually near water, but may be located a considerable distance from bay or lake shore. The nesting geese may actually form a colony with their nests placed a few feet from a sidewalk, a trail, or even a flower bed. While the female sitting on eggs scarcely notices people walking, jogging, and cycling past, the gander stays close by and often protests any approach toward the sitting female. As is usual with most sitting birds, the female leaves her eggs only with great reluctance. When the nesting birds do find it necessary to leave the eggs, they head for a nearby lawn or golf fairway for the green food so popular with geese. A small colony on the shores of Lake Washington, in the state of Washington, numbers as many as twelve nests in a circle measuring approximately one hundred feet around. The Canada goose in the non-breeding season is much more elusive and shy than it is when it incubates eggs.

The white stork of Europe and Africa (*Ciconia ciconia*) has also developed an apparent need for civilization: it commonly uses the chimneys and rooftops of houses in central Europe for nesting. In fact, many human inhabitants in stork nesting areas are unable to use their fireplaces because of the massive nests blocking the chimneys. Nevertheless, these folks regard the presence of the nesting storks as a good omen; they even construct convenient, flat platforms high on their rooftops just for the storks' use.

There are a number of birds that will use, without hesitation, nest boxes built by man if they are placed properly in the correct habitat. It is not difficult to lure a pair of violet-green swallows into a nest box built to the correct specifications. House sparrows and starlings are quick to accept such offerings if properly made and placed. The purple martin (*Progne subis*) will select an "apartment" in a house

constructed for its benefit and placed in an open area in the back yard. Barn swallows, in a way, are accepting manufactured nest structures when they use the rafters, girders, beams, or walls of a house or barn. Indeed, in some parts of its nesting territory the barn swallow is never found using a natural site. Some species of piciforms can also be lured to a nest box nailed to a convenient and suitable tree. Wrens usually build rather large, bulky nests with an entrance on the side, but some wrens use a natural hole such as a deserted woodpecker nest or even a cavity under the eaves around a building. The house wren will use a nest box. As a boy on the farm, I built small nest boxes and nailed them in back of an open knot hole in the barn for the house wren.

Some species of waterfowl, such as the wood duck, find and use natural cavities in dead trees, but a nest box made to the proper dimensions is also accepted by this species. Mergansers can be persuaded to use an artificial nest box, but it must face the open water so that the adults may enter and exit in a straight line. I once took fourteen hooded merganser eggs from a nest box. This large set was probably laid by two merganser hens; there were minor observable

*The brown creeper (*Certhia americana*) willingly builds in a nest box like this attached to a tree. The structure was removed from the tree so that the interior of the nest box could be photographed. One-half life size.*

differences in the eggs of the clutch and it is not unusual for two females to use the same nest. Nor is it uncommon to find the eggs of the wood duck and the hooded merganser in the same nest.

The brown creeper literally spends its life on the trunk of a tree. Its sharp toes and stiff, pointed tail feathers make it well adapted to living on a tree trunk, along with woodpeckers and other piciforms. The creeper uses the space created by a piece of dead bark pulling away from the tree as its nest. It also readily accepts a piece of bark nailed to a tree trunk. There must be entrance holes on each side permitting the bird to enter or exit under the roof. Moss or twigs tightly stuffed in the bottom of the cavity form a foundation. In most cases, a nest structure of this kind is more stable and secure than a natural one.

Bluebirds and several other species utilize natural cavities for their nests. These species readily accept nest boxes placed on trees in suitable habitats. Consequently, various groups of people have developed programs to reverse the decline of the western bluebird (*Sialia mexicana*). In suitable areas, hundreds of nest boxes have been introduced into regions acceptable to these birds. Up to 86 percent of these nests can be successful, and total fledglings can number in the hundreds. Building plans for nest boxes for various species of birds are readily available from a number of public libraries.

A few birds place their nests in habitats that appear to be greatly out of place. It doesn't seem natural for a duck to place its eggs high in the hollow of a dead tree; yet this is the place from which the small, precocial, downy young of the wood duck, dried within a few hours after emerging from the egg, must somehow make it to the ground. Nor does it seem in character for the graceful, long-legged heron, perfectly at home wading in stream, lake, or salt water, to build its nest high on the limb of a forest tree many feet from the ground. And it is surprising to find the nests and eggs of some ground-nesting passerines placed in a tin can on a rubbish heap at the edge of the prairie. But unlikely or not, one never ceases to marvel at the instinctive ingenuity that leads birds to place their eggs in remarkably successful locations.

NEST CONSTRUCTION AND MATERIALS

Nest building appears to be largely instinctive. Although the young may be taught to fly, to defend themselves, and to feed, they are at no time taught how to build a nest or what materials to use. Yet for many generations birds have continued to build nests so characteristic of their species that an ornithologist can name the bird by looking at the nest. Precocial birds are in the nest as chicks for only a few hours at most, leaving very little time for learning. Even altricial young, which spend days or months in the nest, are obviously not using their time learning the rudiments of nest architecture and construction. They are at best seeing and touching only the interior of the structure. And yet they repeat the nest building techniques of their ancestors almost exactly. The downy young of the wood duck (*Aix sponsa*), for example, scrambles upward on the inside of the nest cavity and jumps out to the ground below. When the duck becomes an adult and starts its first reproductive cycle, instinct leads it back to an acceptable tree to use a hollow cavity for a nest.

The marsh wren seems to have a particularly strong instinct for nest building. It builds a series of nests, one of which will be used for the containment of eggs, while the others function only as decoys to confuse potential predators. It is possible that the marsh wren does this not only to perplex predators, but also because its urge to build nests is unusually strong. Perhaps, after a nest is built, the urge to lay has not kept pace with the construction of the nest, so building

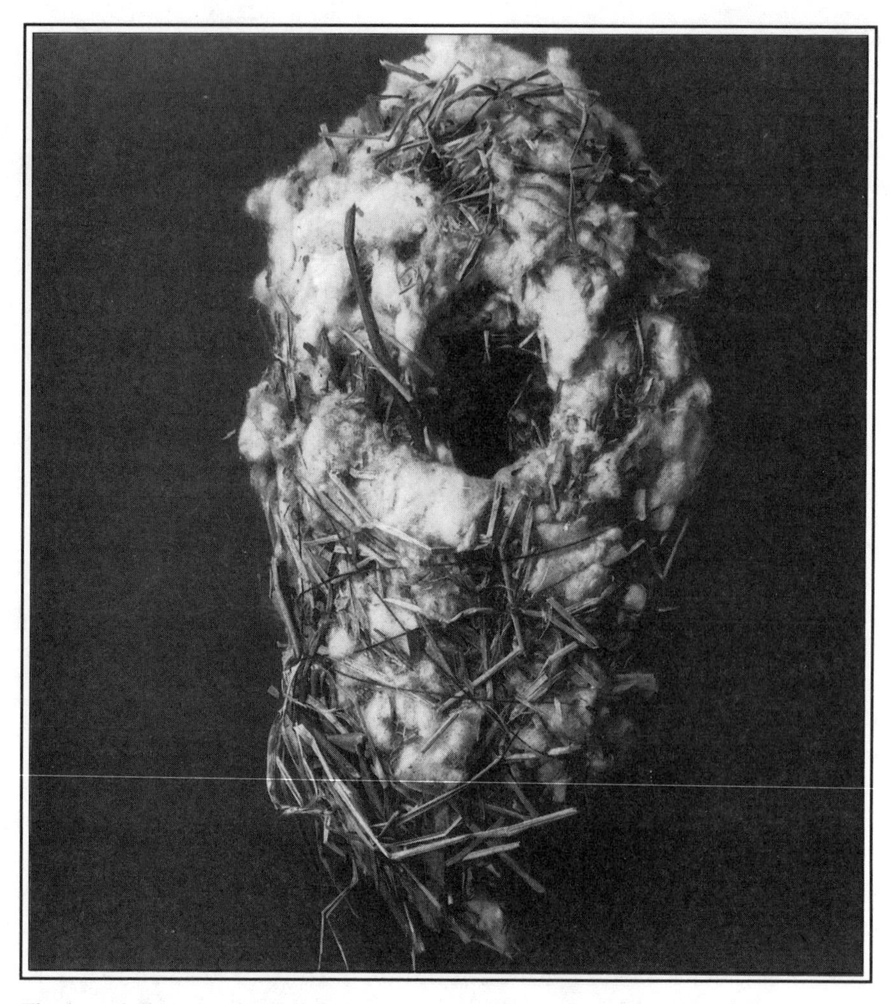

The long-billed marsh wren (Cistothorus palustris) *seeks out a territory in the marsh or swamp. This bird uses the old wet leaves of sedges, rushes, and cattails for its domed nest with a side entrance. This particular nest is a decoy and unlined. One-half life size.*

activities might continue until egg laying begins. The final nest is the only one in the series that is lined for warmth—with cattail down—and it is the one most difficult to locate, tucked away in the marsh vegetation and well hidden from view within the habitat. J. H. Bowles, in 1909, spent one day in prime marsh wren habitat and counted fifty-three nests. Only three held eggs.

Although the nests of birds within a species are strikingly similar generation after generation, among all bird species there is an infinite variety of nest sizes and shapes. The giant platform nest built by the heron or eagle contrasts sharply with the tiny, soft, tightly woven and skillfully built structure of the hummingbird. But each is just right for the successful reproduction of that species.

Nest Shapes

The simplest nest of all may be a shallow, saucer-like scrape in the sand. From the viewpoint of evolution, this nest may have emerged during the courting, dancing, and mating activities of a pair of birds. The next evolutionary step in the construction of the ground nest might be the addition of twigs or other vegetation that naturally folded and compacted into the nest as the female moved over the incubating eggs. Ground-nesting birds may form a scrape in soil or sod and line it loosely or tightly with grass, hair, or other substances.

A domed nest might logically follow if and when the brooding female smoothed the grass around and over her body. Kitchin, on May 4, 1931, collected the nest and eggs of a Hungarian partridge (*Perdix perdix*) at Spanaway, Washington. The female on this nest had covered her back with grasses so that only her head was visible. The bird that retreats into a cavity to build its nest is probably seeking a substitute for the domed structure in the open.

It takes little skill for a bird species to gather materials and place them indiscriminately in a pile. The nests of galliformes and waterfowl are constructed in no particular shape; they must simply hold enough building material to protect the eggs, cover them when necessary, and keep egg and hatchling warm when the adult is absent.

The mound-building megapodes live in the rain forests of New Guinea and Australia, where the forest floor has little sunshine but

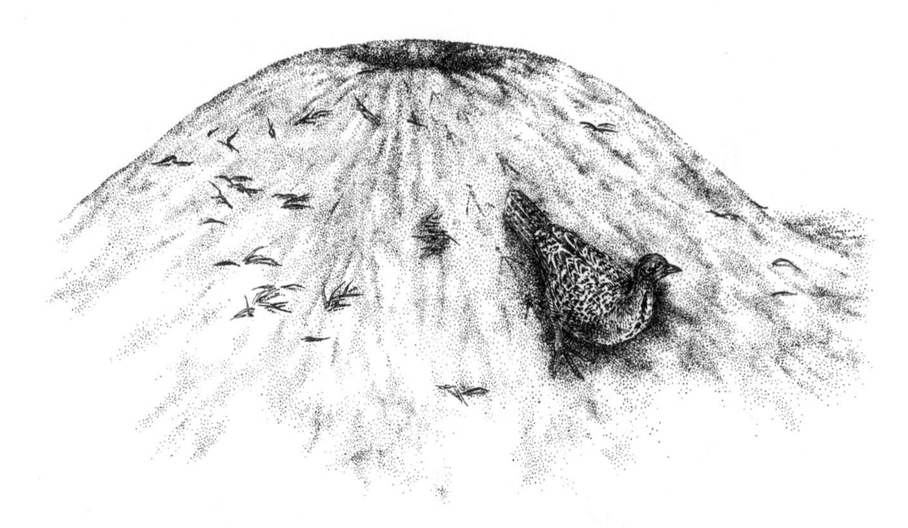

The mound nest of the mallee fowl.

great quantities of rotting vegetation. These birds build up large mounds from this vegetation over a period of many weeks. Including the time expended to build and tend the mounds after the eggs are laid, the reproductive cycle may take as long as ten months. Megapodes excavate the top of the mound and place the eggs inside, then cover them over to utilize the heat given off by the decomposing vegetation.

Many tree-nesting species of birds commonly build nests similar in form and shape but distinct in size, material, and location within the tree. Most of these tree-nesters build nests that can be classified as cups, bowls, domes, saucers, or pensiles. The foundation for these structures usually consists of rather coarse materials such as twigs or rootlets lined with finer, soft substances such as feathers, leaves, grasses, hair, plant down, or wool. Such linings are extremely important because they help retain heat for the developing eggs and the nestlings. Altricial young would suffer and die from the cold unless a thick nest lining provided insulation to retain the heat supplied by the brooding parent.

The duck-feather lining of this yellow-rumped warbler's nest (Dendroica coronata) *helps keep the young warm after they hatch. Life size.*

*The nest of the band-tailed pigeon (*Columba fasciata*), a crude platform of twigs, offers no heat retention for this one-egg clutch. One-half life size.*

The Construction Process

One has to wonder how a bird always manages to construct a nest of exactly the right size. The explanation is quite simple: a bird sits in the spot where the nest structure is to arise and builds the nest around its body. Although this is an easy process for ground-nesting birds, it is not so simple for those that must construct hanging nests, such as vireos, bushtits, and orioles. These builders become very skillful. They wrap one end of a grass rootlet or hair to one branch of a fork of a limb, reach over, and tie the opposite end to the other branch of the fork, being careful to leave enough slack to begin to form the bottom of the cup. When enough of these initial foundation "struts" are in place to support the bird's weight, the bird sits in this skeletal structure and proceeds to complete the nest to fit its body.

The pensile nest of the bushtit, for instance, is built from the top down, with the first structural fibers put in place around the branch to which the swinging structure will be attached; grasses, rootlets, and mosses are woven into the tube or passageway as the nest lengthens downward. The bottom of the nest is first formed into a skeletal framework in which the builder can sit, and then the thick cup at the bottom is woven around the body to the specific size of the bird. The final addition to the nest is the soft lining of plant down, brought in by way of the tiny opening on the upper side of the nest. The lining is taken down to the bottom and packed into position by the feet. The bushtit nest, a marvel of engineering and construction, usually survives the rigors of wind and rain for an entire breeding cycle. In fact, old remains of bushtit nests can be seen hanging on bushes a year or more after their construction.

In contrast, the oriole's pensile structure is open at the top and hangs as a sack. The nest of the vireo, about the size and shape of a teacup, is sturdily built and hangs with its rim attached to a fork in the branch of a tree. Because the vireo nest does not sway with the movements of the branch as do the nests of the oriole and bushtit, the vireo's nest could be called semi-pensile. A further modification of the suspended nest is that of the kinglet. This nest is sack-like and constructed of thick mosses lined with feathers; it is sus-

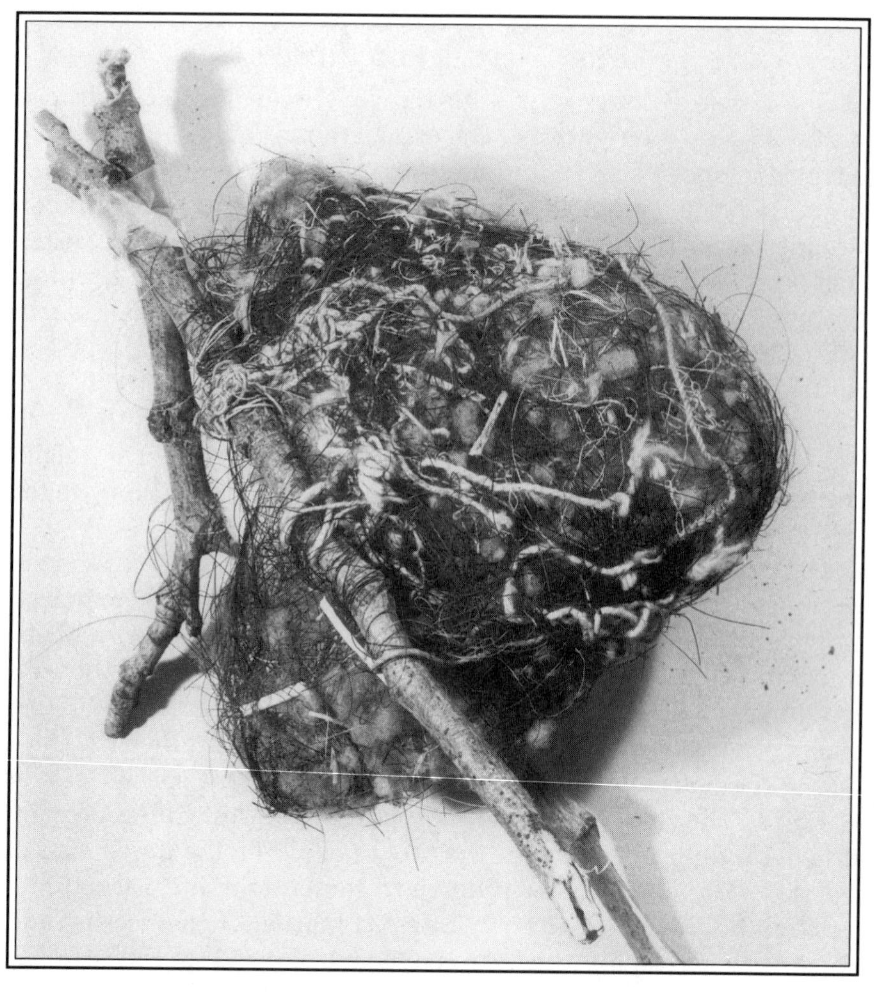

*The nest of the northern oriole (*Icterus galbula*) is constructed almost entirely of horsehair with a scattering of twine and cottonwood fiber. One-half life size.*

A close-up of the oriole nest shows how the nest material is wrapped around the supporting twigs. Life size.

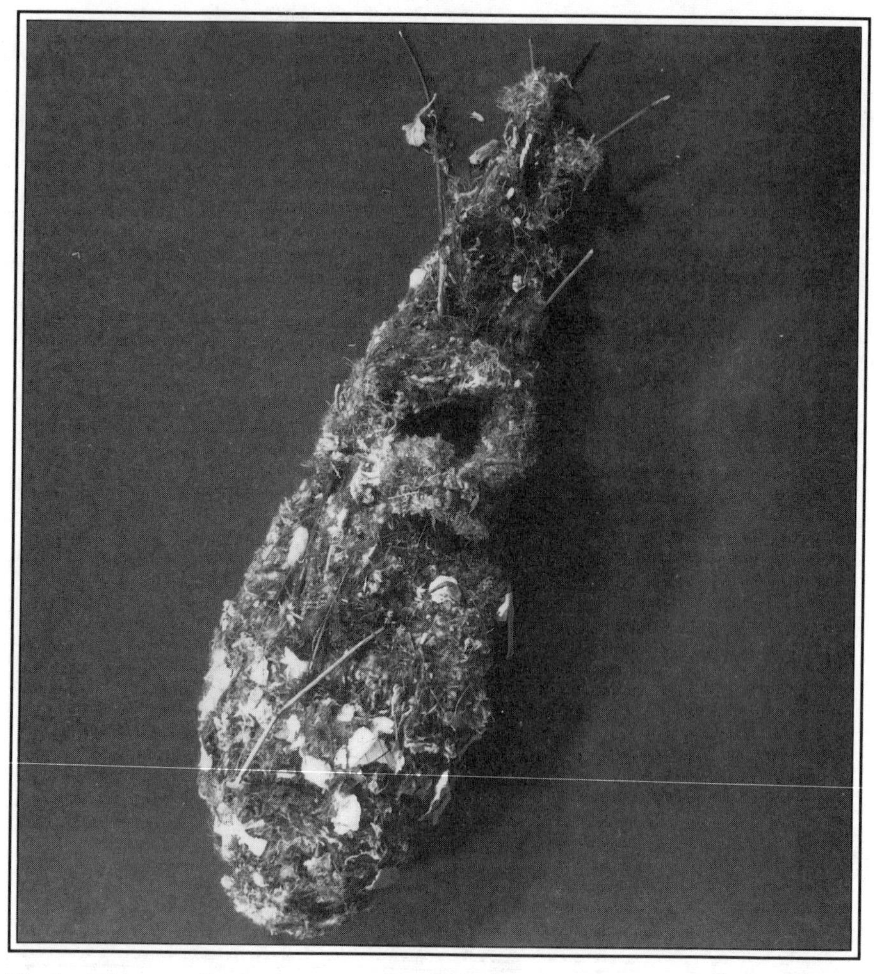

*The bushtit's nest (*Psaltriparus minimus*) is a classic example of pensile construction. One-half life size.*

The yellow warbler usually constructs its cup-shaped nest in the crotch of a tree.

pended from the secondary branches of the large limbs of a tree.

Some of the weaver birds of Africa construct flask-shaped pensile nests from the top down to form two chambers. The entrance to the upstairs compartment is by way of a long tube that attaches the nest to the tree at the point where the elaborate construction process began. This is the only part of the nest attached to the tree. The egg chamber is downstairs; the birds that enter this chamber through the upstairs compartment line it with plant down and fine grasses.

The robin nest, although not pensile, is built with similar techniques, though with some modification. The robin first builds a rough bowl structure to be lined with mud. Adult birds bring the mud, apply it to the lining area, sit in the bowl, and, with backward strokes of their legs and feet, shape and smooth the mud lining. Normally, the adults then leave the unfinished structure long enough to permit the mud to dry. Soon thereafter they line the nest completely with a loose layer of dry grass.

The hummingbird nest is built entirely by the female, who chooses a sloping branch on which to saddle the structure. She attaches the basic material of the tiny cup to the branch with wrappings of spider webbing that is used like binding twine. In some species, an outer covering of lichens is wrapped in place in the same way. Floss or down from seeds and leaves is placed inside the cup and shaped to the bird's body with each addition of lining material. The long bills of many hummingbirds prevent them from placing the lining material around their bodies while they sit. Instead, many find it necessary to stand on the rim and place the lining material inside; the bird is forced to jump in and out of the nest to pack the lining into shape.

The winter wren prefers to place its nest as a lining in a cavity in the soil around the root mass of a fallen tree. But if no such habitat is available, the wren may form a hollow nest (with an entrance on the side) from mosses woven together and tied tightly to the exposed roots or the branches of an available tree. Since such a nest is not supported by surrounding material, the birds weave twigs with their beaks and feet into the moss to form a criss-cross matrix immediately below the side entrance. Thus, a firmer area strengthened by twigs supports their feet when the birds enter or leave the nest, thereby reducing wear and tear on the fragile moss structure.

Many birds in the process of building a nest will stop their building activities if disturbed and desert the nest site. Generally, those species easily driven from the nest area in the early stages of nest building are increasingly reluctant to stop building when the nest is closer to completion. The same general rule can be seen in the egg laying stage. Birds with partial clutches laid in the completed nest will desert more quickly than birds with clutches completed.

Nesting Materials

Birds are quite adept at locating and using materials that have

*This photograph shows the nest of the winter wren (*Troglodytes troglodytes*). Note the matrix of criss-crossed twigs to strengthen the nest at the entrance. Life size.*

Nest of the American dipper (Cinclus mexicanus). *The nest is composed entirely of living moss kept moist by the spray from an adjacent stream. One-half life size.*

been traditionally used by many generations of nest builders. Their selection of nesting territory is based partly on the availability of substances needed to construct the typical nest for the species. Efficiency increases and time and energy are saved if birds, in the process of building nests, do not have to search far for favorite materials.

The dipper (*Cinclus mexicanus*) represents a species that demands an extremely specialized territory for the availability of proper nest materials. The dipper's nest, a great hollow ball of moss, is built along the borders of a stream, not only because of the availability of the living moss but also so that spray from the stream keeps the moss in the nest alive and green. Dipper nests are also built on a

*This western lark sparrow (*Chondestes grammacus*) used an old mockingbird nest. The nest is composed largely of long twigs and covered with blooming poverty weed for camouflage. One-half life size.*

rock in the middle of the stream or even at the back of a waterfall.

Although birds use their favorite natural nesting materials whenever possible, they will also use material that is close at hand. If a nest is placed in a lichen-covered bush, liberal use will probably be made of the available lichens. A nest and set of eggs from the Rivoli hummingbird (*Eugenes fulgens*) (UPS Museum), taken in the

Huachuca Mountains in Cochise County, Arizona, in 1901, show what remarkable use the female can make of available nesting materials. A maple limb was completely covered with foliose lichens for a distance of about eighteen inches. The hummingbird placed its nest in the center of the lichens and constructed it in such a fashion that the lichens covered the outside of the nest completely and blended perfectly with those covering the limb.

In some birds, selection of the material is the duty (or prerogative) of one of the pair. Penguins, for instance, commonly offer stones to the mate in the nesting season. The male marsh wren gathers nearby wet reeds and energetically takes them to the female who is busily weaving the nest. She may or may not accept the offerings, but because the leaves are but a few feet away, any wasted energy by the male appears to be of little importance.

The most abundant natural materials for nests come from the plant kingdom: roots, stems, leaves, mosses, bark, algae, ferns, lichens, grasses, and cottony substances. Other natural substances birds commonly use include mud, feathers, fur and hair from animals, pitch, salivary secretions, and guano.

Many species of birds build their nests of only one material. The golden-crowned kinglet (*Regulus satrapa*) uses nothing but moss. The summer tanager (*Piranga rubra*) uses only pine needles, and even omits a separate lining substance. The black-headed grosbeak (*Pheucticus melanocephalus*) uses only twigs, coarse ones on the outside of the nest but fine ones in the center. Willow bark is a favorite choice of the fox sparrow (*Passerella iliaca*), while grass is exclusively used by the hooded oriole (*Icterus cucullatus*), the glaucous-winged gull (*Larus glaucescens*), and the honeycreeper (*Coereba*). The honeycreeper gathers the flowers of grass and weaves them into the structure along with the grass leaves.

Other species use several readily available nest materials. The pyrrhuloxia (*Cardinalis sinuatus*) uses dry grass and small pieces of bark to form its nest. Various lichen species are used by gnatcatchers, waxwings, and thrushes to protect and camouflage their nests.

Birds are opportunists, and many species—especially passerines—will often use any material at hand, natural or not. A hummingbird will use fine threads from a mop that has accumulated

*The globular nest of the golden-crowned kinglet (*Regulus satrapa*) is completely formed of moss. Approximately life size.*

*The summer tanager (*Piranga rubra*) constructs its nest using little but pine needles. Two-thirds life size.*

*The black-headed grosbeak (*Pheucticus melanocephalus*) uses only twigs to make its nest; there is no special lining. Slightly reduced.*

*The black-billed cuckoo's nest (*Coccyzus erythrophthalmus*) is a rude platform of criss-crossed twigs. This bird is a new-world cuckoo and is not parasitic. Two-thirds life size.*

*The nest of the western tanager (*Piranga ludoviciana*) is composed entirely of rootlets and twigs with a sparse lining of horsehair. The twigs gradually become smaller toward the center of the nest. Two-thirds life size.*

*This nest of the fox sparrow (*Passerella iliaca*) is composed of dark shredded willow bark scantily lined with grass. Life size.*

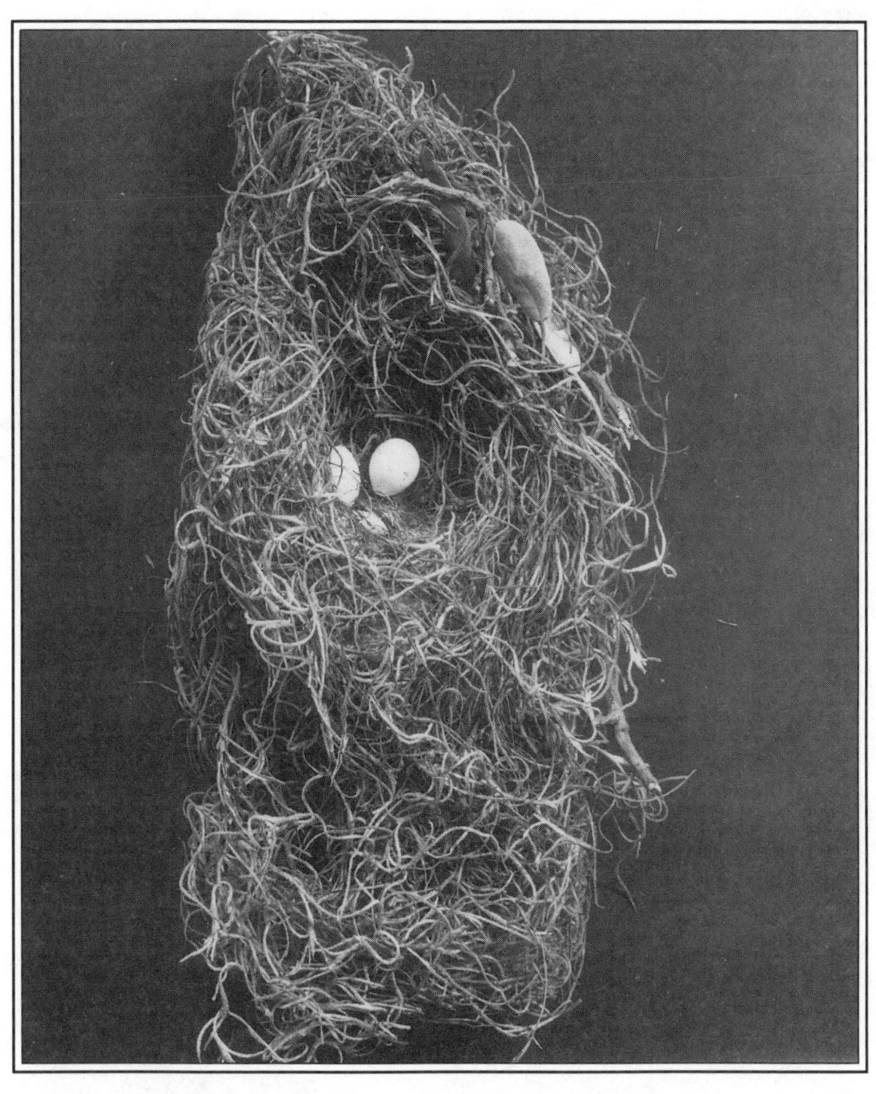

*A Parula warbler's nest (*Parula americana*) suspended in tangled branches of Spanish moss. Two-thirds life size.*

*Nest and eggs of the Bahama honeycreeper (*Coereba flaveola*). The nest is loosely constructed of grass blades including many dried flowers. Life size.*

household dust if the mop is accessible to the hummingbird. The ultimate opportunists might be those species that select any available "tin roof" to shelter their nests, including the piles of tin cans discarded in fields by homeowners before garbage collecting became mandatory. One junco pair used any can with a large enough opening that rolled from the top of the pile and came to rest on its side. The nest, made of grass and mosses, was placed inside the can to provide both warmth and protection. The can concealed the nest and in many cases proved to be rainproof—an avian Quonset hut!

Some species, such as the crested myna (*Acridotheres cristatellus*), western kingbird (*Tyrannus verticalis*), and California shrike (*Lanius ludovicianus*), show an astounding ability to seek out and use materials of any kind available. In addition to twigs, feathers, grasses, and other natural materials, these species and others gather great quantities of string, cloth, rubber, and paper to give bulk and substance to their nests. The ultimate example of this is shown in a recently acquired oriole nest. The parent birds had somehow come upon some monofilament fish line. This line was about a 10-pound test and evidently had been found by the birds in a tangled condition, but there were enough unbroken pieces that they were able to construct the nest almost entirely of this artificial substance. The birds had also found somewhere a few inches of twine and several pieces of plastic tape, evidently from someone's tape recorder. The entire nest had been woven around pendant branches of a birch tree and lined with a few small, wispy pieces of cotton. The monofilament line, which could not be broken easily or affected by the weather, was an excellent choice for building material.

In lieu of twine some birds select natural cobwebs, both to include in the nest and to use in attaching it to the supporting branches. The egg cases of certain spiders may be placed on the outer walls of the nests of some birds. These give the appearance of petals from the flowers of adjacent shrubs.

It is not unusual to find the shed skin of a snake built into a nest. While it serves no structural function, it probably frightens the marauding predator aiming to rob a nest of its eggs.

A horsehair lining is popular with the chipping sparrow (*Spizella*

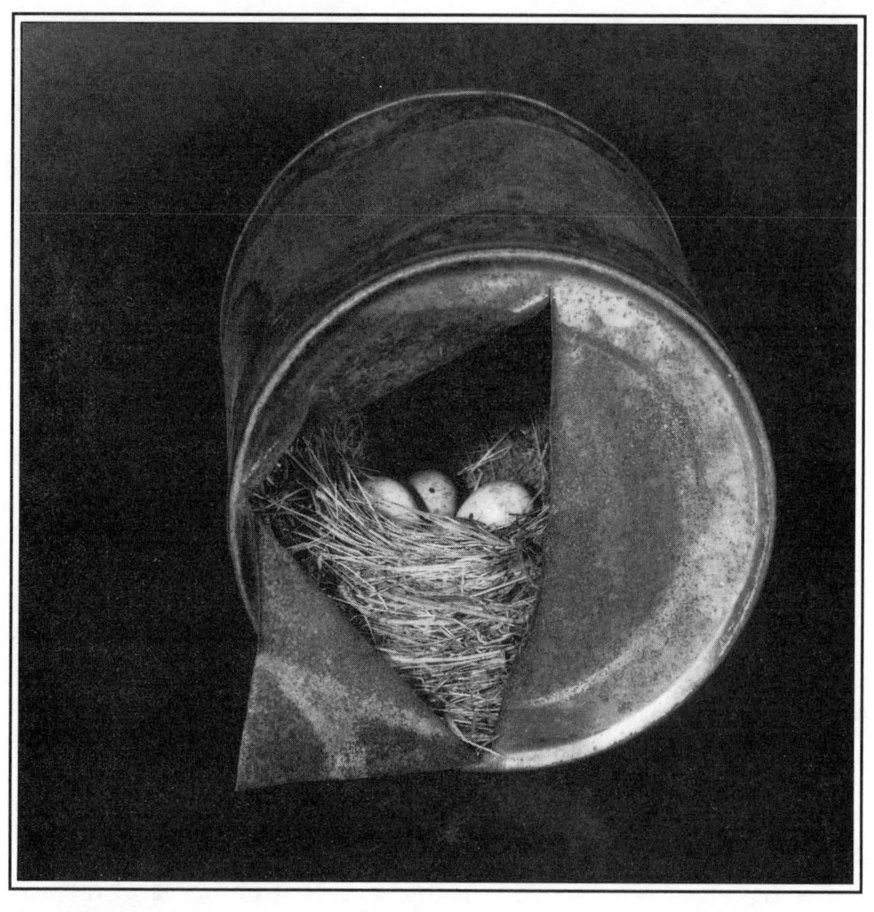

*This dark-eyed junco set (*Junco hyemalis*) was collected in 1931 before the days of the modern can opener. Two-thirds life size.*

The nesting burrows of the rhinocerous auklet honeycomb the bank on a headland adjacent to the sea.

A rhinocerous auklet in the grass at the entrance to its burrow.

The island home of the rhinocerous auklet, tufted puffin, and cormorant.

Newly hatched goslings of the Canada goose, less than two days old.

A female Canada goose at its nest with a typical clutch of six eggs. Placed only a few feet from the water's edge, the nest is built of cattail leaves mixed with down feathers. The clean eggs and fresh appearance of the down and plants denote the beginning of incubation.

In this nest the eggs are soiled and the down looks worn, which indicates that the eggs have been incubated for many hours.

Nest and eggs of the great blue heron. This nest was 180 feet high in a tree.

A nesting colony of great blue herons. These nests are often sixty feet or more above the ground.

The shy bittern (above) tends to dwell in the bog or marshland in the thick vegetation.

Heron habitat is more open — not as thickly vegetated as that suited to the lifestyle of the reclusive bittern.

Nest colony of the cliff swallow. The mud nests in a large colony may number in the hundreds.

Cliff swallows gathering mud for nest building.

A newly hatched tern chick with three unhatched eggs. Terns usually have a clutch of three; additional eggs in the nest are probably laid by a different female.

A one-hour old Caspian tern chick. The prominent egg tooth at the tip of the beak will disappear after several days. Note the color differences between this photo and the previous one. There is considerable variation in the color of the down within a species, just as there is variation in eggshell pigmentation within a clutch.

Competition on the nesting grounds: tern eggs eaten by gulls.

Island nesting colonies of the Caspian tern. Terns, like many seabirds, nest on islands where possible. These colonies provide a degree of safety from predators.

The nest of the Caspian tern is a mere scrape in the sand. The vegetation is not part of the nest.

Coastal rocks house seabirds such as murres, puffins, cormorants, and petrels.

Eggs and newly hatched chicks of the cormorant. Cormorant young are entirely black and have very little down.

*The comb-crested jacana (*Irediparra gallinacea*) is polyandrous. In times of danger the male carries its young tucked under its wings. The long-toed feet hang down from under the parent's wings.* Photo by T. Mace

The killdeer, like most shorebirds, lays four protectively-colored eggs in a small, shallow ground nest.

A killdeer displaying the "broken wing" behavior to lure a predator away from eggs or young. Photo by T. Mace

The long-billed marsh wren builds its nest of cattail leaves in a habitat of swamp vegetation.

Weaver birds in Africa build thousands of nests to form enormous colonies. The nests hang on the ends of limbs, out of reach of predators.

Nesting burrow of the fairy penguin taken on Phillip's Island, Australia.

The Cape Kidnappers gannet colony in New Zealand is said to be the only mainland gannet colony in the world. It has several thousand members.

A large nesting colony of the gannet. Note that the birds in the distance appear to be in rows, and far enough apart that adjacent birds cannot steal the nest materials.

An incubating gannet is sitting on its single egg. Gannets have no brood patch and must cover the eggs with their warm feet.

The nutcracker and gray jay choose nest sites at timberline in the mountains.

Rain forest growth of old trees provides a home for the spotted owl in the Olympic mountains in the northwest corner of Washington state. Logging practices in rain forests are depleting old-growth habitats, and this owl is quickly becoming a rare species.

The tundra of the northland is the nesting habitat for the snowy owl and for many shorebirds.

This ring-billed gull nest was placed deep in wet, marshy vegetation.

Most downy seabird young, like these ring-billed gull chicks, tend to be speckled to aid in concealment from predators. These nestlings are several days old.

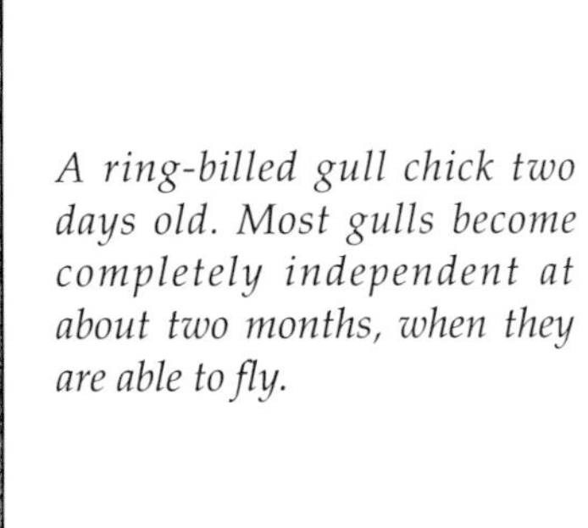

A ring-billed gull chick two days old. Most gulls become completely independent at about two months, when they are able to fly.

One of these glaucous-winged gull eggs is lighter than the other two; sometimes the last egg has less pigment than the first ones laid.

Nest and eggs of the spotted sandpiper on a river island.

Lapwing and its eggs. The eggs are protectively colored, which helps conceal them in an open field. Photo by T. Mace

Nest and eggs of the pipit in the heather of the high mountains.

The superb lyrebird on display. This rear view shows the two lyrate (lyre-shaped) feathers, twleve plumes, and two filiform feathers. Photo by L. H. Smith

The Sherbrooke Forest in Victoria, Australia, is the home of the superb lyrebird (below right). Photo by L. H. Smith

A male lyrebird singing on a mound. This is a typical pose during courtship with the tail held forward over the head. Photo by L. H. Smith

*This nest of the California shrike (*Lanius ludovicianus*) is composed of a great variety of materials. Much of the nest proper is made of cloth and string woven about a scant matrix of twigs. The lining is largely of wool. Three-quarters life size.*

*The nest of the northern mockingbird (*Mimus polyglottos*) is composed of weed stems, willow cotton, and dried flowers. The long spines of the haw aid in protecting the nest and its contents. One-third life size.*

This chipping sparrow's (Spizella passerina) *nest lining is composed entirely of black horsehair; some chipping sparrows build totally white linings. Life size.*

passerina) and the chestnut-sided warbler (*Dendroica pennsylvanica*). The chipping sparrow appears to prefer either white or black hair because only one color is usually found in a nest. Flycatchers are fond of using the hair or fur of various mammals for their nests. Collectors have found nests of the ash-throated flycatcher (*Myiarchus cinerascens*) that were made entirely of the soft hair of four-footed animals, including cows, rabbits, and skunks. In sheep country, great quantities of wool that snags on shrubs when sheep pass by is picked up by many birds and incorporated into their nests.

Spectacled eider (Somateria fischeri) *nest with an abundance of black down. One-half life size.*

*Nest and eggs of the Bohemian waxwing (*Bombycilla garrulus*). The nest is composed of a matrix of spruce twigs with liberal use of a black lichen (*Allectoria*); the lining is largely made of wool with some grass. Life size.*

*This compact nest of the Swainson's thrush (*Catharus ustulatus*) is composed largely of a foliose lichen (*Usnea*) and some twigs with a lining of shredded inner bark from the cedar tree. One-half life size.*

If there are feathers of one kind or another in the chicken yard, swallows are quick to pick them up and use them in nest building. Ducks and geese commonly line their nests with down for heat retention and cover the eggs with down while the adult birds are away feeding. The down is plucked from the breasts of the adult birds. Waterfowl that use feathers in their nests are less dependent on other materials such as grass or reeds within the natural habitat.

Shore birds, in general, are not skilled in constructing elaborate nests, but they all take advantage of available grass, moss, lichens of the tundra, and the leaves and stems of aquatic plants. Sandpipers nesting in the tundra of northern regions construct thick nests of mosses and lichens that aid in retaining heat. Long-

*The ovenbird's (*Seiurus aurocapillus*) bulky nest is carelessly made and contains the skeletal remains of leaves from deciduous trees. One-half life size.*

billed curlews, which nest in dry habitats, tend to limit their nest material to a shallow accumulation of dried grasses. Some killdeer nests include not only a sparse amount of vegetative material, but also a small collection of seashells that apparently caught the eye of the adult bird while the nest was under construction.

Leaves are common in the nests of many bird species. The ovenbird (*Seiurus aurocapillus*) weaves the skeletons of old leaves into its

Nest and eggs of the rusty song sparrow (Melospiza melodia). *The nest is completely sheathed in dry leaves. Life size.*

The Louisiana waterthrush (Seiurus motacilla)*, like its relative the warbler, shows a fondness for leaves in its nest. One-half life size.*

nest while the song sparrow (*Melospiza melodia*) covers almost the entire outer surface of its nest with dead leaves. The Louisiana waterthrush (*Seiurus motacilla*) places its nest on top of leaves, which make a good foundation for grass nests, as does its close relative the Kentucky warbler (*Oporornis formosus*).

Mud is another important nest substance. The cliff swallow (*Hirundo pyrrhonota*) locates a convenient mud hole and carries small quantities of mud in its bill, ultimately forming a gourd-shaped chamber. Nesting pairs of cliff swallows may gather more

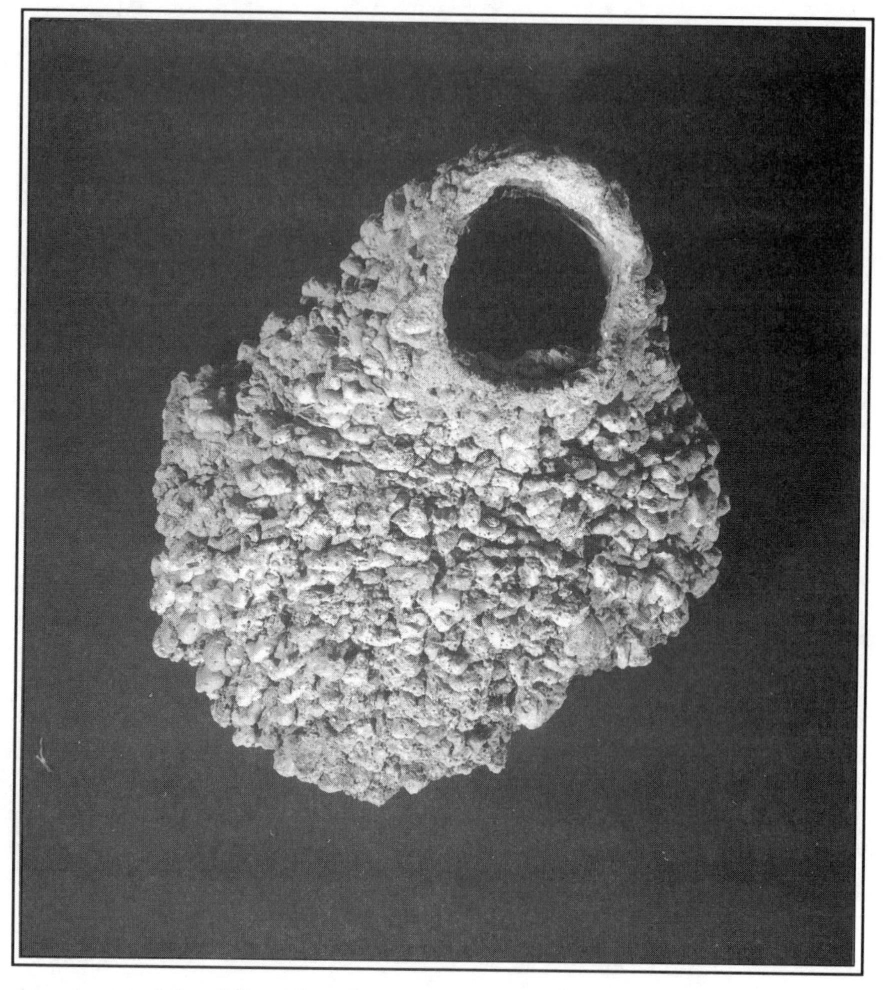

A mud nest of the cliff swallow (Hirundo pyrrhonota)*. More than four hundred blobs of mud are needed to complete the nest. One-half life size.*

than four hundred small blobs of mud to complete their nests. These structures are cemented to cliffs, bridge girders, or the walls of buildings such as large grain elevators, or even beneath the overhanging roofs of houses. The nest opening is on the side of the nest, and its interior is lined with grass. This swallow is highly colonial, and great numbers of nests may form a solid layer of mud huts.

*The black phoebe (*Sayornis nigricans*) also uses mud to build its nest. Life size.*

The mud structure eventually crumbles, but not until the young have left the chamber.

Mud is also the principal nest material used by the black-browed albatross (*Diomedea melanophris*) of the Falkland Islands. This bird's nest looks like a pedestal composed almost entirely of mud. At the top of the structure is a deep, cup-shaped depression, which often looks too small for the incubating bird that sits upon it. These albatrosses are skillful artisans; their structures withstand the wear and tear of an incubation period that can last up to sixty days.

The flamingo (*Phoenicopterus ruber*) builds a large pedestal mud nest up to thirty inches in diameter and saucer-shaped at its top. To perch successfully over the single egg that this species lays, the incubating bird must fold its legs in such a fashion that its heels extend over the outside of the nest. The hammerhead stork (*Scopus umbretta*) of Africa builds a large nest of sticks liberally cemented with mud. A. G. Vrooman, on June 23, 1921, at Santa Cruz, California, found an unusual nest of a black swift (*Cypseloides niger*) in a cavity in the roof of a cavernous ocean cliff ten feet above the breakers. No nesting material was present but an egg was laid inside a ring of wet mud (UPS Museum).

The nest of the barn swallow (*Hirundo rustica*) consists of mud blobs that form a cup, which is then lined with feathers. This nest is placed on a building rafter or over a doorway or other structure. Barn swallow nests are not found in colonies or natural places such as rock walls or the sides of cliffs, although they are sometimes found in caves. The black phoebe (*Sayornis nigricans*) also prefers mud as a building material and constructs a nest similar to the barn swallow's. Phoebe nests are placed on girders below bridges over water or on the beams of deserted buildings. While the swallow lines its nest with feathers, the phoebe prefers grass and rootlets.

Mud is used in a different way by the hornbills (*Buceros*) of Asia and Africa. The nest is in a hollow tree, and the male seals the female inside with mud to prevent her from leaving until the eggs have incubated and the young have departed from the nest. The male brings food and passes it to her through a small hole in the mud seal.

The colonial nesting gannet builds small guano and mud nests raised a few inches off the ground. Partially dried seaweed is a popular component in these nests. While the seaweed apparently serves no structural function, it is usually present and is apparently subject to great demand; it is continually filched by adult birds from adjacent nests. The gannet must expend considerable energy flying to the top of a cliff with a beak full of wet seaweed picked up from the sea far below.

The oil bird (*Steatornis caripensis*) of South America, a member of the Caprimulgiformes, with industrious imagination builds a cone of guano deep in a natural cave. While many birds build their nests of camouflaging material, the Caprimulgiformes themselves are camouflaged: they tend to blend in with the caves where they place their two eggs. The eggs also show a high degree of mimicry: they look like two small rounded stones.

Many birds display a "pack rat" instinct in their incessant drive to collect nesting materials far beyond their structural needs. For instance, the burrowing owl (*Athene cunicularia*) covets horse manure as a constituent of its nest. It is a rare burrow that is without at least a small quantity of horse manure. This bird's instinct to collect a substance that is of no apparent value to its reproductive cycle and is not an integral part of its natural territory cannot be explained. One can speculate, however, that the manure odor serves to mask any nest scent picked up by a predator.

Many species of birds use an astonishing conglomeration of nesting materials. Members of the family Corvidae are especially adept at locating and finding nest material. Examination of more than 150 nests of crows, ravens, magpies, and jays showed that all contained a great variety—sticks, rootlets, mud, leaves, cow dung, bark, bark fibers, rags, cotton, string, moss, lichens, dried ferns, pine needles, horse hair, cottonwood twigs, wool, sheep bones, cattle bones, wire, and feathers. In the bulky nest of the crow—in addition to the normal materials—there may be pieces of lost jewelry that have caught this bird's eye.

It is not unusual to find coarse twigs and grasses forming the bulk of the raven's nest, while its lining includes shreds of bark and pieces of shiny metal and paper that appeal to the raven. Over several days, the raven, almost in a playful fashion, punches holes in any pieces of paper so collected and placed in the nest. Eventually the hole-ridden paper will be discarded. (In addition, pieces of metal and paper provide play objects for the incubating bird—perhaps helping it to while away the boring hours!) In the dry Upper Sonoran zone, a raven nest was once found on the top of a windmill water tower. The eccentric shaft of the whirling windmill forced the vertical shaft to move up and down as water was pumped. The raven constructed its nest so that the edge surrounded this moving part. Sheep ribs served as a matrix for the nest, and their curvature formed its circumference.

The magpie has a domed nest well anchored in a tree. Several feet in diameter with a side entrance, it houses the usual litter of leaves, grass, rootlets, and other available material. These rustic, bulky structures are neither elaborate nor artistic, but they do protect the eggs and young from hungry predators. Often some

*The gray jay (*Perisoreus canadensis*) a montane species, intersperses the heavy, thick body of its nest with wool to add warmth. Two-thirds life size.*

mud is used to aid in cementing the coarse twigs together. The long tail of the magpie sometimes protrudes through the nest opening.

The California jay (*Aphelecoma coerulescens*), unlike most corvids, builds its nest with a simple foundation of twigs and a single type of material for the lining.

Swallows that build in a nest box fill the cavity with whatever material is available, as do house sparrows; feathers are favorite objects for filling the cavity. A small cup is formed in one corner to

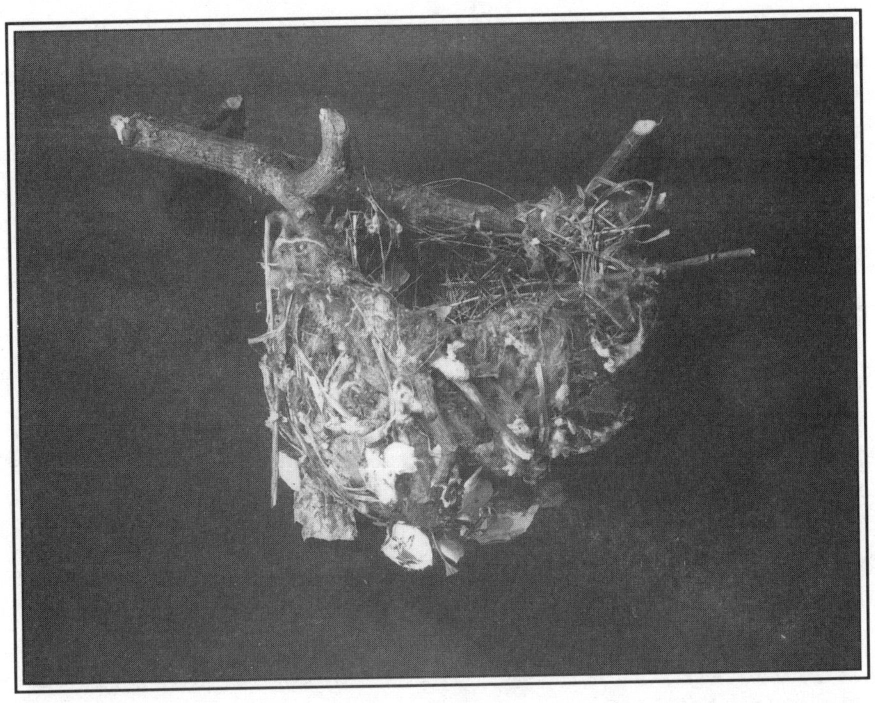

*Nest of the solitary vireo (*Vireo solitarius*). The white materials are egg cases of spiders. One-half life size.*

house the eggs. House wrens fill their nest cavities with twigs. Horned larks use moss from an adjacent prairie and place nests alongside a small rock (perhaps to keep the nest a little warmer), while meadowlarks use the available vegetation to build a cup-shaped structure beneath a clump of prairie grasses to take advantage of a "living roof." This dome makes the nest less visible. Additionally, when the parent meadowlark returns to eggs or young it lands a few feet away from the nest and approaches it via a circuitous route that twists through small tunnels in vegetation, thus eluding or confusing a possible predator.

The vireos are the artists of nest building. The solitary vireo (*Vireo solitarius*) selects a wooded area and hangs its nest from a small, forked limb of a tree. Its nest is the size and shape of a teacup. It is compact, tightly woven, and smoothly lined with bits of grass—where available—and the sporophyte generations of certain mosses. In one species the exterior of the cup is extravagantly covered with white egg cases of spiders. The egg cases fluttering in the wind look like a cluster of white flower blossoms, perhaps helping to conceal the nest. It must take considerable diligence on the part of the vireo to search out the white egg cases to cover its nest.

Reusing a Nest

Routinely, students interested in ornithology ask whether or not a pair of birds uses the same nest year after year. A related question that often arises is whether individual birds use the same nest for two or more broods in one season. Some species do indeed use the same nest, especially if it is built of sticks intertwined in such a way that the nest becomes very strong and durable. It may take only a minimum of repair work to make such a perennial nest usable.

Smaller birds that do not construct large, strong nests able to survive the rigors of winter must annually build a completely new nest. For example, the hummingbird may build a new nest on the remains of the nest constructed the year before. But it is comparatively rare for a pair of birds to construct one nest per season and hatch two or more broods in that nest. Nevertheless, J. H. Bowles observed a single pair of western robins that used the same nest to raise three broods in a single season. A first set produced four young. A second set brought forth three young. Some weeks later a new lining of dead grass was added, and later two more young hatched. Bowles believed that no mud was added before the new grass lining, but he did deem it probable that the bird must have bathed and, with wet feathers, dampened the original mud lining so that the new grass lining could be placed to the satisfaction of the female. While there is no proof that these three sets were the result of at least the same female, circumstantial evidence seems to point to this as a fact. The male used the same singing perch throughout the production of the three broods, and the female became so tame that it was necessary to lift her off the third brood in order to examine it. This was in contrast to her behavior during Bowles's visits to the first brood: she was noisy, wild, and distraught.

Eagles, ospreys, and other raptors commonly use the same nest

HURLEY COLLECTION

OOLOGICAL SPECIMENS

FROM COLORADO.

EGGS ARE MARKED
A.O.U. No 337a / 1/3 SET INDEX / NO. EGGS IN SET

Eggs of the Krider's Hawk: "Buteo borealis kriderii"

Date of Collection and Locality.
May 22: 1888 On Crow Creek. 21 miles East of Greeley. Weld County. Colorado:

Incubation fresh: and slight.

Nest 45 feet from ground in a dense cottonwood well made of coarse sticks, with a lining of grasses and raw cotton. It is the Custom of our Hawks on the Plains to pick from the Cottonwood tree the grape like clusters of the cotton ball and put a few of these in their nests, which also bursts under the incubation and make a soft raw cotton lining. (over)

This Set is mentioned especially in Davies Nest and Eggs 5th Edition page 208:
Oliver Davie sent it to G. H Messenger Linden Iowa:
This is also the Set (and the bird) on which I got credit for the addition of the "Krider's Hawk to "The" Colorado List of Birds" By Prof W. W. Cooke.
Re written at Linden Iowa. October 4-1905
Fredrick M. Dille

On this egg card from 1888, Dille notes the custom of hawks on the plains to use cotton from the cottonwood tree in their nests.

every year but add new sticks. An eagle nest may weigh up to two tons. Many raptors line their large, shallow nests with twigs and leaves. Davie cites an example in a nest record of Krider's red-tailed hawk (*Buteo jamaicensis*) collected by Dille in 1888 in Weld County, Colorado, near Greeley. Dille notes, "It is the custom of our hawks. . . to pick from the cottonwood trees the grape-like clusters of the cotton ball and put a few of these in their nests, which also bursts [sic] under the incubation and make a soft, raw cotton lining" (UPS Museum). Another nest—this one built by a red-shouldered hawk (*Buteo lineatus*) and recorded in California on May 19, 1909, by A. B. Howell—contained a considerable variety of cotton from the seeds of the cottonwood tree where the nest was found (UPS Museum). Cotton balls make warm, soft nest liners.

EGGS

AND

EGG LAYING

The selection of the nest site and the construction of the nest must parallel the development of the egg within the female, because the nest must be available at the time of egg laying. Most birds place their eggs in the nest a short time after the structure is complete. There are, however, some exceptions to this general rule. For example, the robin delays several days, allowing its mud-lined nest to dry before placing a grass lining over the mud. Only then can the eggs be laid.

Timing the Eggs

The ovarian follicle may liberate a new ovum a few hours after an egg is laid, but rarely is more than one egg laid in twenty-four hours due to biological processes that take time to build a new egg. In clutches of more than one, an egg is usually added to the set every other day. The two eggs in the clutch of the Adelie penguin (*Pygoscelis adeliae*) are consistently laid one and one-half days apart. Those rare ducks that have two functional ovaries can lay two eggs per day.

The time of day that eggs are laid cannot be related to the order or family of birds. In clutches of two or more eggs the time of laying might be related to time of maturation of the egg within the oviduct of the species. Most eggs are laid at about daybreak. The time of laying may be partly explained by the fact that the reduced activity of the female during the night could concentrate more metabolic activity on egg maturation and laying. If there is a disturbance at the nest that prevents a female from laying an egg at dawn and this egg is laid only several hours later, the egg has started to divide while still in the oviduct.

Egg-laying patterns according to calendar time are quite exact in many birds and consistent with complete life cycles. Migratory species nesting in northern climates do not reach nesting grounds until late spring or early summer because their food, which consists of insects, seeds, fruit, and worms, is available only after winter's cold has given way to the warm summer season. For both altricial and precocial birds, time permits only one brood per season in these northern climates. A few exceptions exist. Polyandric phalaropes in the tundra produce two or more clutches, each concurrently incubated by a male. Sanderlings commonly produce two clutches, each in a separate nest and each incubated by one parent of the pair.

The nighthawk (*Chordeiles minor*) of the Northwest, a late migrant, does not lay its eggs until the first of July, and young are not hatched until early August. In this species the young can be fed well into autumn; there is no time for a second brood. The typically long nest cycles of many altricial species prevent them from completing more than one successful reproductive cycle per season.

In general, only birds that lay early in the season in more temperate climates may have time to produce a second set of eggs. These second broods, when established, may only amount to 10 percent or less of the total population hatched in the spring. The second egg clutch in many species is probably laid to replace eggs lost or destroyed in the first clutch. This theory is not completely tenable, however, because the second batch of eggs may appear not immediately but rather after a lapse of several weeks. In some species many late eggs never hatch, and many of the young that do hatch do not survive, probably due to autumn's cold.

Seasonal periods for egg laying are remarkably constant. The short-tailed shearwater (*Puffinus tenuirostris*) in the Southern Hemisphere consistently lays its eggs within a two-day period each year. In northwestern America the Clark's nutcracker (*Nucifraga columbiana*) and the gray jay (*Perisoreus canadensis*), both montane species, lay their eggs in early spring while the ground is still covered with snow. The young then hatch in early summer and have

Many gallinaceous birds, because they are precocial, have large clutches. This California quail (Callipepla californica) *has a set of eighteen. Two-thirds life size.*

time to fledge before the short summer gives way to cool autumn. At lower elevations, the dipper and the killdeer usually lay their eggs by mid-March. The marsh wren (*Cistothorus palustris*) and the Virginia rail consistently complete egg clutches by early April. Wherever the species nests, it is most efficient to confine reproductive activities to as short a period as possible; some anatomical and physiological changes related to reproduction would not be beneficial to the species if they persisted throughout the year. The enlargement of the gonads is a case in point. If these structures were permanently rather than periodically enlarged, carrying them within the body would result in a great loss of energy.

Clutch Size

As a general rule, altricial birds tend to have small sets of eggs while precocial birds generally produce large eggs and large clutch-

*A nest of twenty eggs of the redhead (*Aythya americana*). Measurements and egg shell textures suggest that two females were involved in the production of this large clutch. One-half life size.*

es. A high degree of consistency in clutch size is observed in those species that lay no more than four eggs to a set. Gallinaceous birds and waterfowl generally have large egg sets that commonly number from six to ten but can number as many as twenty. An exceptional clutch of a redhead (*Aythya americana*) was taken by D. E. Brown on May 27, 1931, in eastern Washington on the shores of Moses Lake (UPS Museum). This set of twenty-eight eggs occupied an area much too big for one hen to incubate successfully. It probably represented the efforts of at least two females. An unusual clutch of hooded merganser eggs (*Lophodytes cucullatus*) was probably produced by two hens. This can be demonstrated by slight differences in egg measurements.

Seabirds such as terns and gulls usually produce no more than three eggs. Hummingbirds lay two; nighthawks, two; eagles and hawks, commonly no more than three or four; and falcons, three or four but occasionally as many as five or six. There is a remarkable consistency in the clutch numbers of many shore birds. Knots, sandpipers, turnstones, dowitchers, and some plovers lay four eggs. Almost the only exceptions in this large group are the oystercatchers (*Haematopus* sp.) and snowy plovers (*Charadrius alexandrinus*), whose clutches usually number three.

Specific research regarding clutch size in the reproductive cycle of the Caspian tern (*Sterna caspia*) showed irregular sets of four or five eggs in some clutches instead of the usual three. In a large colony of Caspian terns, researchers color coded the eggs in nests, each with a normal clutch of three. This first egg was marked with a green felt pen. The research team visited the colony daily over a period of about a week in order to determine the nature of the laying cycle. Second eggs in marked nests were coded using a blue marker. The third egg was color coded red. Subsequent visits to the colony as incubation progressed and egg laying ceased showed many nests containing four or five eggs. Close examination disclosed duplicate colors on the eggs in nests containing more than the normal three. Whether the host female stole the eggs or a donor gave them away was not determined.

Early naturalists noted a tendency for other terns to have clutches of inconsistent sizes. Two sets of eggs produced by a least tern (*Sterna antillarum*) and taken by J. H. Bowles on an island off Cape Cod, Massachusetts, on July 25, 1893, each contained four eggs rather than the normal two (UPS Museum). Bowles speculated that each set of four was probably laid by two females. Similarly, W. P. Nicholson, on June 27, 1926, collected a set of four least tern eggs in Brevard County, Florida (UPS Museum), and noted that it was the only set of four on record at that time in the state.

Often the first egg ever laid by an individual bird is much smaller than normal and sterile. This extra egg is regularly observed in first clutches of the domestic chicken, the cormorant (*Phalacrocorax* sp.), and other birds. Consistency in clutch size appears to be established in an individual female after her first laying season. The first egg laid by a young female may appear before the physiological processes in egg laying have developed in their correct sequence. This explanation, while not exact, gains credence with ornithologists since first eggs usually contain no yolk.

The exact percentage of infertile eggs produced by wild birds is not known, but is quite low. Occasionally an infertile egg is present in a wild clutch, but it is rare to find an entire infertile clutch. A set of the eastern bobwhite (*Colinus virginianus*) collected by Walter B. Savary at East Sandwich, Massachusetts, on June 8, 1932, proved to be infertile when taken and artificially incubated (UPS Museum). But researchers should be careful when drawing conclusions concerning the failure of an egg to hatch, because factors other than fertility can cause the development of the egg to cease.

Why and how adult birds recognize the completion of the clutch, which signals the onset of brooding behavior, is unknown. Removal or destruction of all or part of the clutch will postpone brooding and stimulate the laying of a replacement set. Some species, such as petrels, lay only one egg per season; they cannot replace this egg if it is destroyed. But many species respond to an extreme loss of eggs by laying eggs continuously—producing far beyond the original clutch numbers. A. C. Bent states that Phillips took almost one egg each day, for a total of seventy-one eggs over seventy-three days, from the nest of a northern flicker (*Colaptes auratus*). He always left one egg there as a nest egg.

Egg replacement has become a fine art in man's manipulation of the domestic chicken's clutch size. An individual leghorn chicken may be induced to lay well over three hundred eggs per year. One theory suggests that brooding behavior in birds is not triggered until the feel of the eggs to the adult setting bird is normal. Perhaps that normal sensation against the incubating bird's skin induces cessation of ovulation.

Ornithologists working in aviaries have determined that if the

Abandoned nest and two eggs of the cedar waxwing (Bombycilla cedrorum). *A tiny egg in a clutch is usually the first egg ever laid by the individual female.*

first clutch (consisting of one egg) of the California condor (*Gymnogyps californianus*) is removed from the nest, the bird will replace it with a second egg. If this in turn is removed the female will lay a third. These three eggs can then be incubated in the aviary with a net gain of two eggs from each female per season. Such a scheme takes advantage of a behavior pattern that is common in many species of birds. It is especially valuable with the condor because this species, which does not mature until it is six years old, is now considered endangered and is perhaps approaching extinction in the United States.

Birds that reach reproductive age late and lay small clutches are usually long-lived and may nest every other year. This is probably nature's way to insure a better chance of continuing a viable population of a species.

Egg Size

Of considerable interest to ornithologists is the study of egg size and egg weight compared to the weight of the adult. Most large, highly altricial birds produce notoriously small eggs compared to the weight of the adult birds. The wandering albatross (*Diomedea exulans*) averages about seventeen pounds, while its egg, measuring 13.5 by 8 centimeters (5.25 by 3.13 inches), weighs approximately one pound, or six percent of the body weight. The California condor is North America's largest bird, with an average weight of more than twenty pounds. Yet its egg, measuring 11.5 by 6 centimeters (4 3/8 by 2 1/2 inches) and weighing six-tenths of a pound, constitutes about three percent of the parent's weight. Both albatross and condor have a clutch of one.

Some birds have eggs that do not follow the usual characteristics differentiating precocial and altricial species. For example, the eggs of the kiwi and the ostrich are laid in small clutches. Yet both these birds are identified as precocial because the newly hatched young are quite capable of skittering over the ground and picking up food for themselves. The young also have an ample coating of down feathers. The large egg, the long incubation period, the presence of down, and the fairly independent young point to the classification of these species as precocial. Only the small clutch is definitely altricial in nature.

The egg of the kiwi (*Apteryx australis*) is 14 centimeters (5.5 inches) long and weighs about one pound, which is twenty-five percent of the body weight of the adult bird. This egg is larger in proportion to adult weight than that of any modern bird. The clutch varies from one to two. This small clutch might have been an adaptation to the special environment in New Zealand where there are no natural ground predators. (Flightlessness in the kiwi may have developed for the same reason.)

The adult ostrich (*Struthio camelus*) weighs about three hundred pounds, but its egg constitutes less than two percent of the parent bird's weight. Adaptive evolutionary processes (through natural

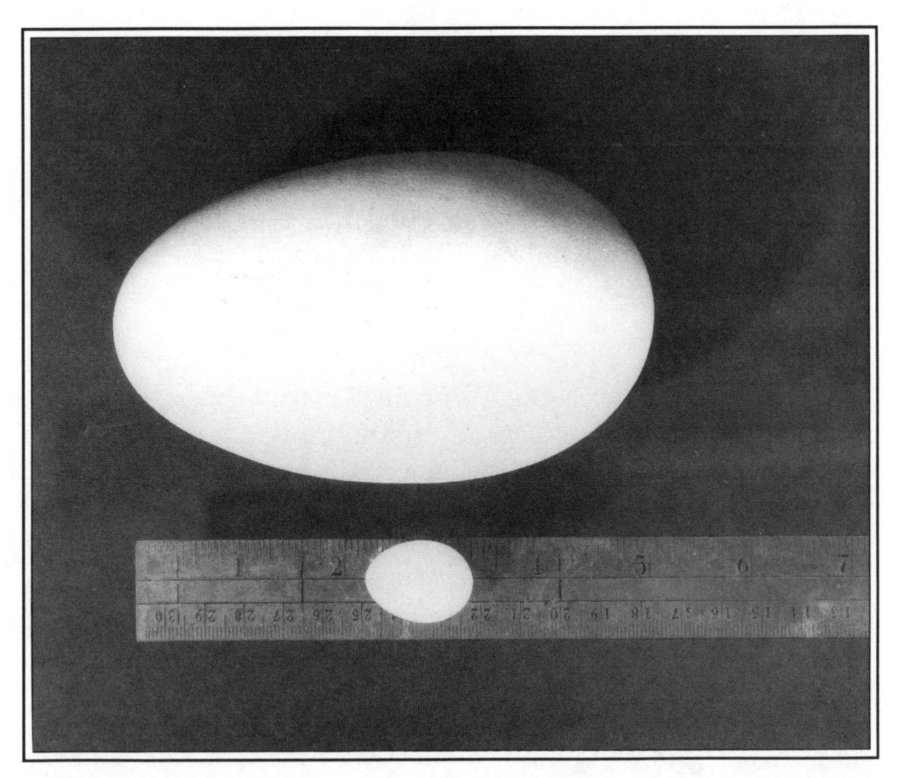

The small petrel and the large albatross belong to the same order (Procellariiformes), but the volume of the albatross egg is 130 times that of the petrel egg. Slightly over one-half life size.

selection) have apparently determined that the size of the ostrich's egg is sufficient to give the young bird a good start toward successful development. The ostrich has a clutch of four or five.

Incubation periods, as will be mentioned later, are definitely related to egg measurements and clutch size. The largest known egg belonged to the extinct elephant bird (*Aepyornis maximus*). Presumably, this giant, flightless bird became extinct between the thirteenth and fifteenth centuries. The egg of this species measured about thirty-eight centimeters (fifteen inches) long and had a capacity of more than two gallons. It weighed more than twenty-five pounds. By comparison, the ostrich egg—largest of modern birds—measures about eighteen centimeters (seven inches) long and weighs approximately three pounds. The egg of the elephant bird would hold the contents of seven ostrich eggs, which is equivalent to about two hundred chicken eggs—or thirty thousand hummingbird eggs. The egg of the rufous hummingbird is only about twelve millimeters (one-half inch) long and weighs approximately one-half gram (one fifty-sixth of an ounce). Eggs of different species of similar habits and habitats within an order show remarkable consistencies in shape and color but vary greatly in size.

Egg Shapes

Eggs can be broadly classified into three major shapes: oval, elliptical, and pyriform. Oval eggs are round on both ends, but have widest measurement above the "equator." Most birds have oval eggs, though in some species, as in owls, the two ends are so nearly alike that the shape tends to be rounded. Eggs of the yellow-headed blackbird (*Xanthocephalus xanthocephalus*) are long ovals.

In elliptical eggs, the curvatures of each end are about the same. Elliptical eggs are commonly produced by some raptors and certain birds with streamlined bodies such as grebes and hummingbirds.

Pyriform or pear-shaped eggs illustrate the extreme: one large end and a smaller end, with the region between the "equator" and the small end almost linear, thus radically changing the oval appearance. Pyriform eggs are characteristic of most shore birds and many passerines. The pyriform egg of the murre rolls in a tight circle on the bare rock of a narrow ledge and is therefore less likely to fall into the sea. The large pyriform eggs of precocious shore birds are placed in the nest with the small end pointing down, so that the incubating parent sits on the larger end. If the eggs were placed horizontally in the nest, the small incubating adult could not cover the clutch.

There is a remarkable consistency in the shape and size of eggs within a species and within a single individual's clutch. Occasionally, however, the volume of one egg may be normal but its shape greatly different, or an egg might be characterized by both greater volume and an unusual shape. An egg in one clutch of a white-breasted nuthatch (*Sitta carolinensis*) measured 24 by 13.5 millimeters (UPS Museum). The remaining eggs averaged 20 by 14 millimeters. In this example the anomalous egg was radically different from the

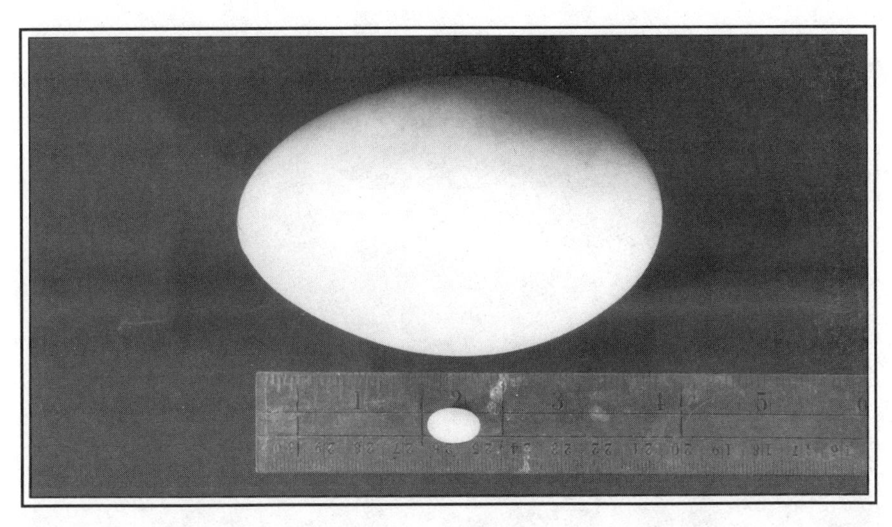

The large oval swan egg is two thousand times the volume of the elliptical egg of the hummingbird, our smallest bird. One-half life size.

The egg of the altricial bald eagle (Haliaeetus leucocephalus) *compared with the smaller egg of the precocial mallard duck* (Anas platyrhynchos). *Life size.*

*The eggs of the yellow-headed blackbird (*Xanthocephalus xanthocephalus*) are long ovals. Two-thirds life size.*

others in both shape and volume. The volume of the aberrant egg was approximately fourteen percent greater than those of normal oval shape.

Eggshells

Eggshells may have a single color or they may possess a solid background color with various markings of different colors. Eggs may be white, pale blue, dark blue, greenish, tan, or brown. Some piciforms produce eggs that are yellowish or orange; this color, however, is not derived from shell pigments but comes from the yolk, which shows through the thin, white shells of some of these species. No truly black eggshell is known. Egg color may vary considerably within one species. The beautiful, glossy eggs of the tina-

*Thick-billed murre eggs (*Uria lomvia*) are pyriform in shape. These particular eggs also exhibit a variety of pigment patterns. Three-quarters life size.*

mous of Central and South America are produced in various solid colors, such as green, blue, purple, wine red, and yellow, but shades of dark brown tend to dominate. All have hard shells with a high "varnish" gloss.

The markings on bird eggs can also vary in size and color. Brown markings can appear on a perfectly white background or on a dark ground. The patterns can also include curiously shaped lines.

Nest and pyriform eggs of Wilson's phalarope (Phalaropus tricolor) *in the tundra grasses. Life size.*

Since the shell is secreted in the oviduct, this is where pigments are applied as the egg descends. The pigments in the blood, which can be chemically isolated and analyzed, are composed of various salts produced by the liver. Often, one or two eggs in a clutch may be visibly lighter in color than the remainder of the egg set. A lighter egg is usually but not always the first one laid in the set because pigment cells in the oviduct have not yet developed. Speed of descent through the oviduct or length of time in the uterus can account for the shape of markings and quantity of pigment deposited in the background color or in patterns. Many eggs show a ring of markings in the area of greatest diameter. It is probable that this pattern occurs when the egg pauses in its descent long enough for pigment to be deposited in the place where maximum pressure,

*This pyriform egg of the pallas murre (*Uria lomvia*) has a pigment pattern of curiously shaped lines. No two eggs have identical markings. Two-thirds life size.*

due to peristalsis, is exerted against the egg by the oviduct.

Colors and patterns of eggshells probably aid the process of incubation and help conceal eggs that are laid in the open. A pigmented egg tends to be warmer because it absorbs certain rays of sunlight. Pigmented eggs with irregular markings are also somewhat camouflaged from predators: the irregular splotches break up the contour of the egg.

Most cavity-dwelling birds lay white eggs, with a few exceptions. Nuthatches and chickadees lay colored eggs even though they inhabit the darkened chambers of dead trees. One theory suggests that these birds used to build nests more exposed to light but have through the generations modified their cycle by moving into the dark for a nest chamber, while at the same time retaining the egg color and markings of their ancestors. The color on their eggs is of no particular advantage inside the tree; therefore loss of color in chickadees and nuthatches is not of evolutionary selective value. Similarly, the rhinoceros auklet (*Cerorhinca monocerata*), which incubates in dark burrows on sea cliffs, has faint marks on its egg, leading to the speculation that this species moved from an open-air nesting habitat and became adapted to the darkness of the burrow for the purpose of egg laying. Even though the auklet makes no nest and the marks are not of highly adaptive value in the dark, they have not completely disappeared. W. H. Dawson, in Dawson and Bowles, says of the rhinoceros auklet:

> "We have curious evidence in the coloring of the egg. Viewed in the large, the purpose of pigmentation is protective. The egg of the gull, exposed to the full glare of day, is dark-colored and so splashed and blotched with brownish blacks that it blends in admirably with its surroundings of dead grasses and dun rocks, and is thus lost to hostile view. But when a species begins to forsake the open and there is no longer need of heavy pigmentation, the egg tends to revert to the primitive white; that is, to unpigmented calcium carbonate. Now in the case of the Rhinocerous Auklet's egg, we find traces of an ancient color pattern, undoubtedly heavy, still persisting in faint lines of umber and in subdued shell markings or undertints of lavender and lilac.
>
> These to the ornithologist are eloquent of a time ages ago before the race went moon-mad."

In general, large eggs must have thicker shells for strength to support the weight of heavier egg contents. But shell thicknesses vary considerably in fresh eggs of comparable sizes regardless of avian relationships or similarities of species. For example, the fresh shell of the large egg of the ostrich is over two millimeters thick and is very hard. The egg of the hooded merganser, which is also very hard, measures approximately three-quarters of a millimeter thick, about twice the thickness of the egg produced by the wood duck, and heavier—even though the eggs of both these ducks are comparable in size. Precocial eggs tend to be larger and thicker than altricial eggs.

Hatching success depends in good measure on the thickness of

A black guillemot egg (Cepphus grylle)*. Notice the rings of pigment around the large end. Life size.*

the eggshell. The tensile strength of the shell must be great enough to protect the egg contents when the egg is rolled, bumped, stepped on, or otherwise mechanically moved about in the nest. At the same time, eggs cannot be so rigid at hatching that the young find it difficult or even impossible to emerge. The tensile strength of the shell diminishes as incubation progresses because much of the calcium carbonate is removed by the embryo and incorporated into the developing bones. Modern-day pesticides that accumulate in the bloodstream and tissues of the adult female inhibit deposition of calcium carbonate on the egg while in the oviduct, thereby weakening the shell. Such eggs rarely hatch; the soft shells are broken before incubation is complete.

BORROWING A NEST AND HOST: BROOD PARASITISM

Brood parasitism is a very unusual and highly modified reproductive strategy found in some families of birds. This behavior is apparently the result of a series of chemical, structural, and social changes. Birds that practice brood parasitism lay their eggs in a host nest and depend upon host parents to incubate and feed the invader until it leaves the nest.

Ornithologists recognize two general types of parasitism: obligate and non-obligate. Obligate parasites always require other birds to incubate all their eggs and care for the young throughout the fledgling stage. Non-obligate parasites normally incubate their own eggs, but occasionally or quite regularly lay eggs in a host nest. Obligate parasitism is found in most members of the cuckoo family, some ducks, all honeyguides, most cowbirds, widow weavers, and some troupials. Non-obligate behavior is observed in starlings, some sparrows, some grebes, Virginia rails, and, quite commonly, among hole-nesting ducks. Hooded merganser eggs and wood duck eggs, for instance, can often be found in the same nest. The eggs of the ground-nesting California quail, bobwhite, meadowlark, and pheasant are occasionally found in each other's nests. Brood parasitism is also the probable explanation for a number of recorded instances where pintail (*Anas acuta*) eggs were found along with eggs of the redhead.

An unusual case of possible parasitism was displayed by a clutch of eggs collected by E. A. Kitchin near Tacoma in Pierce County, Washington, on April

*Nest and eggs of the western meadowlark (*Sturnella neglecta*) with one egg (the large one) of the California quail (*Callipepla californica*). One-half life size.*

24, 1936 (UPS Museum). The nest contained ten mallard eggs and four ring-necked pheasant eggs. The mallard probably functioned as the incubating bird, as suggested by the considerable amount of down that was present in the nest. Storrs Lyman, on April 25, 1921, collected a clutch containing twelve eggs: eight were blue grouse (*Dendragapus obscurus*) and four were ring-necked pheasant (*Phasianus colchicus*) (UPS Museum). The four pheasant eggs, which exhibited subtle variations in color, measurements, and size, were probably produced by two hens.

Parasitic behavior can be a considerable advantage to a species. It relieves the bird of the responsibilities of a normal parent, mak-

This blue grouse nest (Dendragapus obscurus) *contains four eggs of the ring-necked pheasant* (Phasianus colchicus). *One-half life size.*

ing it possible for the parasite to lay a greater number of eggs. Reliable estimates put the number of eggs laid by an individual old-world cuckoo—an obligate parasite—as high as thirty per season. (New-world cuckoos are not parasitic and lay eggs in their own nests, a rudely constructed platform of sticks.)

The habits of the cuckoo were known to the ancients. Perhaps the first substantial written record concerning the behavior of the cuckoo comes from the second-century writings of the Latin author Aelian (quoted in Barrett). Aelian claimed that the cuckoo always laid her eggs in the nests of other birds because she was "too indolent" to take care of her own offspring. Darwin, in his *Origin of the Species,*

A young cuckoo throws the host egg out of the nest. This behavior insures that the cuckoo will get all the attention of the host parent.

said the behavior of the cuckoo was due to variation and natural selection, which he said resulted in the bird's "horrible perfection."

Efforts by ornithologists to explain parasitic behavior have not always resulted in tenable theories. One suggestion is that parasitism originated in the species that typically laid clutches in the abandoned nests of other species. Owls, for example, use the abandoned nest of hawks; starlings and swallows use old flicker holes. A more plausible theory suggests that all of the various physiological changes in a bird do not exactly coincide with the normal progression of steps in the reproductive cycle; thus a species might find it necessary to lay its eggs before a particular hormone level has liberated the instinct for nest building. The cowbird caught in such a quandary, for instance, must seek out the nest of another species in which to deposit her eggs.

In other words, the hormone-induced drives that dictate the sequential segments in a normal cycle are altered, effectively eliminating courtship and broodiness in adult obligate parasites. Most cowbirds display a total lack of these behaviors. The shiny cowbird (*Molothrus bonariensis*) of South America does pick up nest materials, but never fashions a usable nest: if the female is unable to locate adequate host nests, she wastes eggs by laying them on the ground.

The adult practice of putting an egg in a host nest is not the only adaptive behavior in parasitic birds: very remarkable changes are observed in the parasitic hatchling. The cuckoo, as young as ten hours out of the egg, develops a very sensitive, shallow depression across its shoulder area. Naked and still blind, it works its shoulders under a host egg or newly hatched young and tosses it overboard. Welty calls this instinct "one of the wonders of the animal kingdom." This behavior disappears when, at three to four days old, the bird has the nest all to itself. The African honeyguide (*Indicator* sp.) develops a sharp bill hook from its egg tooth with which it attacks the host nestling. In some species both mandibles are hooked. After about two weeks, the hooks separate from the bill and fall off. All honeyguides develop calloused heel pads, which give the bird firm leverage when it is hurling the host nestling over the edge of the nest. In some honeyguides, the mandibular hooks in the young are apparently not essential, because the adult female punctures the host eggs when she lays her egg in the host nest. Either way, if all goes according to plan, the honeyguide hatchling should find itself alone in the nest.

The instincts for caring for young are so strong in certain host adults that they often become confused, tossing their own eggs out of the nest as a reaction to a "foreign" object. Not all host birds are instinctively willing to foster a parasitic egg, however. Friedmann, discussing the cowbird, states that host species may be classified as rejectors or acceptors. The robin, cedar waxwing, blue jay, brown thrasher, Northern oriole, gray catbird, and eastern kingbird tend to be rejectors of cowbird eggs. Among acceptor species are the red-winged blackbird, goldfinches, the barn swallow, and the mourning dove.

Old-world cuckoos tend to lay mimetic eggs that mimic those of

*The tiny nest and eggs of Bell's vireo (*Vireo belli*) with two cowbird eggs. Even though the cowbird eggs are easily distinguishable from their own eggs, the adult vireos treat them no differently. Two-thirds life size.*

*This warbling vireo (*Vireo gilvus*) nest was suspended from a wire fence and contains one egg of the cowbird. Two-thirds life size.*

other species; there are many striking examples of color patterns on cuckoo eggs strongly resembling the color pattern and markings of the host egg. Individual females may lay eggs in a variety of colors. In some cuckoo species the female tends to lay blue eggs in a nest whose host eggs are blue. The same phenomenon has been observed with brownish cuckoo eggs, which often are found in host nests with brownish eggs.

Cuckoo eggs without any markings are often placed in the nests of hosts whose eggs are unmarked. Honeyguide eggs are also white and possess no markings. Host species of the honeyguide are usually hole nesters that lay white eggs in dark places, such as the Piciformes that nest in trees and other birds that nest in burrows. Cowbird eggs, on the other hand, are non-mimetic, which demonstrates that certain acceptor species are not influenced by egg color and pattern (though rejector species may be able to identify and thus reject a foreign egg by pattern).

Success in fledging a parasite species is dependent upon a number of factors. The parasite egg must be laid in the host nest with the greatest speed and least disturbance to the host. Egg laying usually occurs on a warm afternoon—when the host is most likely to be away from the nest. Cowbirds can slip into a nest, lay an egg, and depart within a few seconds. Some instinct tells the female cowbird to place its egg in a host nest containing eggs that approximate the size of cowbird eggs. While the cowbird usually lays only one egg in a host nest, it has been known to lay two or three. Host eggs may number four or more, though clutch size of the host seems to be relatively unimportant to either host or parasite. Additionally, the cowbird usually chooses a host species with an incubation period of approximately the same duration as the cowbird's. Cowbird eggs have so frequently been observed hatching a day or two earlier than host eggs, it must be more than coincidence. This timing obviously gives the young cowbird an advantage over host young.

It is also best if the food habits of host and parasite are similar. Cowbirds seem to regularly place eggs into the nests of altricial species like themselves, although there are exceptions. Friedmann lists a cowbird egg found in the nest of a Virginia rail, a precocial

*Another host, this time a northern waterthrush (*Seiurus noveboracensis*), accepts a cowbird egg. Both species' eggs are very similar in size and markings. Three-quarters life size.*

*A Kirtland's warbler nest (*Dendroica kirtlandii*) with two cowbird eggs. Three-quarters life size.*

species whose cycle is not at all comparable to the cycle of its parasite. Another rare example finds a cowbird egg placed in the nest of a killdeer, a highly precocial species. These few examples could either be abortive and purely accidental, or they could represent a desperate attempt to lay an egg when no suitable host nest is available. Passerine birds with altricial nest cycles similar to that of the cowbird, such as flycatchers, thrushes, bluebirds, warblers, and waxwings, are commonly parasitized, although some are classified as rejector species. Most of these species construct open, cuplike nests; this makes it easy for the parasite to lay its egg and to destroy host eggs. The female cuckoo, for instance, often takes one host egg and either swallows it or flies off with it. Sometimes, however, the construction of the host nest makes it impossible to remove an egg. In the domed nests of some species, the entrance is so small that the parasite cannot even descend into the interior of the nest. In those cases, a cloacal insertion by the female cuckoo is necessary to force the egg, with the help of gravity, into the bottom of the closed structure. Examples of cowbirds laying eggs in closed host nests are rare because the cowbird has no cloacal structure capable of inserting eggs into a host nest. In the domed pensile nests of the bushtit and the nests of some wrens, the cowbird parasite sometimes tries to enlarge the entrance, usually with little success.

In all cases, brood parasitism seems to clearly benefit the parasite. Although parasitic eggs may be lost when rejector hosts destroy them or when their presence induces the host to abandon the entire clutch, the acceptor hosts more than make up any loss to the parasitic species. On the other hand, the effects of parasitism on certain host species can be highly damaging to their reproductive cycles. Kirtland's warbler (*Dendroica kirtlandii*), a rare and endangered species in the United States, nests in a restricted habitat in Michigan's jack pine forests. In recent years, 60 to 70 percent of Kirtland's warblers have been parasitized by cowbirds (Mayfield). The result has been a decreased production of young warblers; this could represent a devastating loss to the population of the species, which numbered only about four hundred at the beginning of the 1972 season. For now the decline is under control.

INCUBATION

A variety of factors, most of them probably hormonal, determines the end of the egg-laying period and the beginning of incubation. Incubation in birds is a process roughly equivalent to pregnancy in mammals. Therefore, in order to insure the reproductive success of birds it is very important that incubation proceeds without major interruptions. In many species both parents are actively involved in incubation. In those cases, the wear and tear of incubation is shared by both parents; should some crisis appear in the incubative process, they can start the cycle over again. The advantage of strong pair bonds is obvious in cases where large clutches are the rule. Here both parents share the incubation and feeding chores.

Some species where both sexes incubate include cormorants, gannets, herons, storks, auks, toucans, woodpeckers, and flickers. Petrels change shifts every seven days. The female sandgrouse incubates by day and the male incubates by night. Pigeons reverse this schedule: the female sits at night and the male takes the day shift. (This reversal is especially striking because sand grouse and pigeons belong to the same order of birds—Columbiformes.)

While incubation is shared by both male and female in many species of birds, there are modifications in this general practice. For example, for most hummingbirds the presence of the brightly colored male would divulge the nest location; consequently, the male avoids the vicinity of the nest during the incubative period. The female, then, is left with the entire responsibility for incubation. The instinct for determining nest location is so strong in hummingbirds that the female also selects the nest site and constructs the nest. In the polyandric phalarope, which nests in the tundra, the male assumes the responsibility for incubation so that several sets of eggs can be produced in a season, each set to be incubated by a male. (The male also develops a brood patch.) Thus more than one set per year is insured even in the short season of the northland. For the same reason, the sanderling (*Calidris alba*) commonly develops two broods. The eggs of the first set are incubated by the male while the second clutch, laid four to six days later, is incubated by the female. The clutches must be incubated more or less concurrently because if incubations were consecutive, the short summer season and the early onset of the cold in the northern climates would wreak havoc on the second brood.

An ostrich nest will usually contain eggs laid by several females. Although this phenomenon is not exactly an example of brood parasitism, it does result in one female incubating eggs laid by another female of the same species. Incubation is carried on by the dominant female ostrich, who incubates by day, and her mate, who incubates by night. The female, less colorful than the male, is partly concealed by her coloration, which yields better protection during daylight hours. The slender-billed shearwater (*Puffinus tenuirostris*) has developed a unique incubation phase. In this species the male is responsible for the first two weeks of incubation; the female takes over for the duration of the cycle, which lasts a total of about fifty-five days. It is difficult to point out any particular advantage to this unusual behavior.

In a number of species of birds the male takes care of incubation entirely. This is true of kiwis, tinamous, some penguins, and phalaropes. In even more species, however, the sole responsibility for incubation is assumed by the female: most puffins, hornbills, flycatchers, larks, swallows, crows, titmice, creepers, wrens, waxwings, vireos, bower birds, and others. The female hornbill mentioned in Chapter 5 must of necessity take care of all incubation. She is fed by the male through a small hole left in the mud entrance seal of the nest.

Temperature Control

Eggs in the process of incubation must be kept warm. The temperatures of adult birds are quite constant, usually varying within a few degrees above 100°F. In the bodies of some passerines the blood may reach a temperature as high as 112°F, while in the kiwi (probably the most primitive of any modern bird) the temperature is a quite constant 100°F. The temperature of the domestic pigeon varies between 105°F and 108°F. In general, the temperatures of large birds tend to be in the lower ranges; smaller birds tend to have higher temperatures. The temperature of the female's blood is usually somewhat higher during the egg and incubation phases in most birds. In polyandrous species, blood temperature in the male is slightly higher during incubation.

It is essential that heat from the incubating adult be passed along to the egg through bare skin or some other body part touching the egg directly. Feathers are insulative and protect the body of the adult from heat loss; if feathers come between the adult's body and the egg, development of the embryo will likely cease. To insure direct contact of egg and parent skin, many species develop brood patches. These are regions on the abdominal wall where feathers either do not develop or are removed by the brooding adult. The brood patch usually appears in the female, except when incubation is largely a function of the male. Then, in birds such as the tinamous and phalaropes, the brood patch is present in the male. Brood patches are liberally supplied with blood vessels and lymph spaces and are somewhat warmer than the normal skin temperature. They are permanent in many species of birds, but temporary in waterfowl; a permanent one would result in excessive heat loss in a species that spends much of its life swimming in cold water.

Some birds possess no natural brood patch and do not develop one during the incubation cycle. In these cases other structures and processes operate during incubation. Waterfowl retain heat around the eggs by surrounding them with down feathers plucked from the adult female's breast. The gannet (*Sula* sp.) drapes its feet over the egg during incubation. The feet of the gannet are large, webbed, and warm enough to insure normal incubation processes in the egg. Gannet chicks often hatch on top of the feet of the parent.

While success and length of incubation depend upon sufficient heat during the entire process, eggs that become too warm are as much a concern as eggs that get too cold. The common nighthawk sits on the ground close to its nest during the heat of the day, holding a wing to cast shade over its eggs. The adult noticeably "follows the sun" and keeps the eggs in the cooler shadow when heat is most intense. Several species of birds must endure unusual heat with low humidity. In Africa the sand grouse (*Pterocles* sp.) routinely brings water back to the nest in its wetted feathers to sprinkle over the eggs as a cooling agent or to supply drinking water to the young. Also in Africa, the sand plover (*Charadrius* sp.) covers the eggs with sand when it leaves the nest, then sprinkles water

Some species of penguins incubate their eggs on top of their feet. This is the best way to keep them warm in the frigid antarctic.

The common nighthawk holds a wing over its eggs to protect them from the midday heat.

over the site to lower the temperature a few degrees.

Special mention should be made here of the unusual incubation processes of the galliform megapodes, the group of birds comprised of the brush turkeys and mallee fowl. They are widespread from Australia to the Philippines, throughout New Guinea, and in adjacent islands in Southeast Asia. The megapodes demonstrate perhaps the most complex incubative behavior of all birds, as they do not utilize their own body heat for incubation. Among this unusual group, several species lay their eggs in warm sand and then abandon them completely. The male of some megapode species tests the temperature of the hole he digs in the sand prior to egg laying. Heat from the sun or occasionally from volcanic steam supplies the necessary warmth through hatching and fledging.

The mallee fowl (*Leipoa ocellata*) regulates its nest temperature by adjusting the amount of fermenting plant material in the mound covering the eggs. The parent birds apparently use their tongues as thermometers. Upon touching the nest material with their tongues or picking up earth in their beaks, a warning of too cold or too hot is conveyed through heat sensors on the tip of the tongue. If the adult bird senses that too much heat is being generated for the welfare of the developing embryos, some of the plant covering will be scratched away, permitting the eggs to cool a few degrees. The reverse is true if the temperature drops: the bird buries the eggs a little deeper for warmth. Some species of megapodes are capable of holding the temperature at the nest in the mound at a remarkably constant temperature of close to 92°F. The megapodes are all highly precocial, and their young remain in the mound until they are able to dig their way to the outside; they begin their independent lives by flying immediately from the surface of the incubation mound. According to Clark, frequent testing of mound temperature by adults keeps the mound material loose, which probably helps the young chick escape when it digs to the surface fully fledged. Within a day after appearing from the nest mound the young are eating and flying. They have no parental care or training and, indeed, never see their parents.

Prehatching Behavior in Adult Birds

Adult birds, in addition to providing constant heat to the developing eggs, must provide other care for them. One essential function is the routine turning of the egg during the incubation process. In the avian egg, the yolk floats on the top of the albumen or white. The air cell in the egg tends to localize toward its large end. It is critical during incubation for the contents of the egg to be free of any permanent contact with the shell. When a developing egg is not moved during incubation, the egg membranes tend to adhere to the shell, thus disrupting the normal development of the embryo. Turning the eggs in the nest also aids even distribution of heat. It is a routine practice for birds to rotate their eggs about 180 degrees each time the adult returns to the nest after an absence for some activity such as feeding. During periods of cold or rain, when a parent bird must cover the eggs for a long period of time without interruption, many species will stand in the nest, reach down, and use

Egg turning during incubation is crucial for the survival of the embryo. It distributes the parent's body warmth to all areas of the egg and prevents the egg membranes from sticking to the shell.

their bills to turn the eggs. Many birds turn their eggs as often as once every few minutes. The egg of the palm swift of Africa (*Cypsiurus parvus*) turns automatically: this bird glues its eggs on the surface of a palm leaf, which then sways in the wind, providing a substitute for the usual practice of rolling the eggs. Commercial hatcheries find that their most successful hatch rate for the chicken is achieved when the eggs are turned every hour. Incubating birds stop turning their eggs a few days prior to hatching.

Although birds pay particular attention to turning their eggs, many are careless during incubation. Caspian terns, for example, appear oblivious to an egg that has rolled a few inches away from the nest scrape and has become chilled. They continue to incubate the remaining eggs. Also, a tern egg is often ignored when it is allowed to become partly or entirely covered with windblown sand. A colony of cormorants (*Phalacrocorax* sp.) is in a constant state of confusion and disarray. Many eggs that are laid in the cormorant nest are allowed to roll out and become lost or be devoured by marauding gulls. Sometimes brooding herons do not tend their eggs very carefully and a few are pushed out of the nest by active young hatched a day or two earlier.

The behavior pattern of adult birds tends to change as incubation proceeds. As mentioned before, many birds will desert a half-built nest or a nest containing one or two fresh eggs of a larger clutch. Even when the clutch is in the early stages of incubation, waterfowl, gallinaceous birds, and shore birds, when disturbed, are quick to slip away from the nest. But as a clutch of eggs advances in incubation the birds tend to stay closer to the nest, and it becomes more difficult to cause the parents to desert their well-incubated clutch. As incubation progresses, the instinct of the adults to remain close becomes more intense until, in some cases, it is almost impossible to cause the desertion without real physical damage to the eggs or the immediate surroundings. The solitary vireo and the killdeer, for example, can even be touched while they sit on a set in advanced incubation; sometimes the female will permit herself to be lifted off the eggs. G. L. Cook, at Tanglefoot Lake in Alberta, Canada, on June 10, 1923, discovered the nest of a bufflehead (*Bucephala albeola*) in a dry, rotten stump. The female bufflehead is a very close sitter. To get her off the nest, Cook actually had to lift her off and throw her up in the air (UPS Museum). In the nestling stage, parents desert the young only after maximum disturbance. Adult birds with young in the nest will show strong anti-predator behavior, even staying to fight a marauding predator until the nest is destroyed or the young are consumed.

Egg Changes During Incubation

During incubation, eggs continuously lose weight due in part to the evaporation of water and loss of respiratory gases through the porous shell, and in part to the metabolism of fats in the egg yolk. This yolk fat represents a much larger percentage of egg contents in precocial species than it does in altricial birds. Hence, the drop in weight of the egg due to fat metabolism is comparatively greater in precocial species than it is in altricial birds. Weight lost during incubation may be ten percent or more of the original weight of the egg.

In addition to physiological and embryological changes within the egg, there are visible external changes. As incubation progresses, eggs lose their bright, more or less clean surfaces and become darker and often quite soiled due to contact with the parent. Shells without pigment often lose their sheen and show internal shadows because of changes in the size of the air cell and rearrangement of membranes. Because of calcium extracted from the shell to establish the skeletal system of the embryo, the eggshell becomes measurably thinner, somewhat translucent, and, in many cases, more fragile.

Period of Incubation

In wild birds, incubation periods within a given species may vary widely, largely due to fluctuations of temperature in the natural environment. When the season is warm, incubation periods are shorter. During such seasons, brooding birds spend considerable time away from the nest. Cooler days cause parent birds to spend more time on the eggs. Eggs laid in the early, cooler part of a season require longer incubation. Within the same species, second clutches need less incubation time during warmer periods. Some second hatchings, however, may show greater attrition if hatching is delayed and cold weather appears early before the young can fly or are protectively feathered. Attrition rates for second-brood Caspian terns *(Sterna caspia)* not fledged by mid-September may be as great as forty percent, due to cold temperatures along the coast of northwestern America.

Some striking examples of variable incubation periods are seen in the heron family (Ardeidae): incubation varies from seventeen days to twenty-eight days. In some hawks (*Accipter* sp.) incubation can range from twenty-eight days to thirty-eight days. Cranes (Gruidae) sit for anywhere from twenty-nine to thirty-six days. In passerine birds, incubation periods are remarkably constant, usually varying no more than about four days.

If temperatures were more constant in the natural environment, variations in incubation periods would be smaller. This can be demonstrated in the domestic chicken. In commercial hatcheries, hatching time can be determined within a period of a few hours when temperatures within the incubator are kept between 36.1°C and 37.8°C (97°F and 100°F).

In many species, the duration of incubation appears to be somewhat related to the size of the birds: small birds have short incubation periods and large birds have longer ones. But there are many exceptions to this supposed rule. The small hummingbird, for example, requires something over two weeks for hatching, while the larger Piciformes may hatch at ten to twelve days. This may, in part, be explained by the nesting habits of the two orders. Piciform species nest in cavities with better protection for the nestlings when they hatch. Hummingbirds, on the other hand, build open, exposed nests with much less protection. The nestlings that have developed longer in the egg are stronger when they hatch, and therefore better able to survive.

In altricial species, there is a rough relationship between the duration of incubation and the amount of time the nestlings spend in the nest after hatching. If the incubation period is long, it is likely that the time spent in the nest will be rather long. Most passerine species (smaller birds) incubate their eggs less than two weeks; a few are shorter, but some, such as crows (*Corvus* sp.), may require three weeks. Other examples of incubation periods show how variable they are: owls incubate for one month; other raptors sit for up to fifty-five days (Welty); the ostrich incubates more than forty days. The eggs of the royal albatross (*Diomedia epomophora*) are incubated for two and one-half months. Precocial birds such as the Galliformes require three weeks for incubation but need no nest time. Ducks incubate their eggs for four weeks. The New Zealand kiwi incubates its eggs for as long as eighty days and leaves the nest with the young immediately after hatching.

Incubation periods are influenced greatly by the latitude in which the eggs are laid. Incubation cycles in northern climates may start as early as March. The great horned owl (*Bubo* sp.) of the North completes its clutch of three to five eggs by the first of April, when the land is still under the spell of cold. The owl must begin incubation early, for it may last as long as a month. By the time the young have achieved their first flight, temperatures are beginning to drop again. The owl is a resident bird (that is, it does not migrate) and is altricial, so it must attend the eggs and nestlings most carefully to prevent them from freezing. Certain migratory species of birds, such as some of the shore birds that use the tundra of northern regions, arrive at their nesting grounds in late spring, incubate and hatch their young in the summer, and are ready to fly south in late summer or early autumn. Although the incubation period of such shore birds may be longer than that of passerine birds, the shore birds are precocious and are ready to fly as early as two weeks after hatching.

Upon the laying of one or more eggs, incubation may begin at once, or it may be delayed. There seems to be no definite rule related to taxonomic groups to explain or predict either pattern. Beginning incubation immediately has both advantages and disadvantages. There is one obvious disadvantage for the medium-sized clutch of five, laid one egg at a time, every other day: the last bird will be at least eight days younger than the first one hatched. But perhaps this is nature's mechanism for providing food where there is a limited supply; the birds that hatch first will be well fed and highly likely to survive. Some raptors, many gulls, hummingbirds, owls, and wading birds do begin immediate incubation. The first birds hatched in the broods of this type begin their feeding as soon as they are able, usually after a few hours. As the subsequent hatchlings appear, the older chicks, stronger and more aggressive, get the lion's share of the food. This may be one reason that most large birds of prey, which normally lay two or three eggs, usually only raise one or two to adulthood.

Cranes lay two eggs at two-day intervals. Despite the small clutch, only one hatchling typically reaches adulthood. For the heron, five eggs constitute a normal clutch. An egg is laid every other day, and incubation starts immediately after the first egg is laid, thus establishing a nesting pattern that may aid in the development of the strongest of the set. Competition for food brought to the nest by adults and carelessly thrown into the midst of the nestlings is fierce; most often the food is gobbled up by the strongest chick. Many adult birds feeding young show a complete indifference when a nestling is pushed overboard in the scramble for food. In some species an adult will even stand by and calmly watch a nestling eat its sibling. Despite the advantage of being hatched a day or more earlier, the first hatchling is sometimes weaker and may be eaten by the second hatched. Perhaps the second hatchling in the nest is genetically stronger than the first.

Whatever the survival value to the species, cannibalism, whether by infanticide or fratricide, is not uncommon among birds. If a young gull or tern strays into a territory other than its own it may be promptly pecked to death and eaten by one of its own species. Within the nest a strong seabird hatchling may deliberately kill its nest mate and eat it on the spot. If a young eagle nestling is too small or weak to seek food actively from a parent, the parent will ignore its position in the nest and feed only the noisy, aggressive nestling. When the weaker nestling dies it is either fed to the remaining young or eaten by the adult. Among some raptors, adults will eat their own young when prey animals are in short supply (though rarely do the parents eat all their young; only the weak).

Gallinaceous birds with large clutches and most passerines begin their incubation cycles only after the last egg in the clutch has been laid. Consequently, all eggs hatch at the same time, and no individuals have any advantages over the others. Progressive hatching would not suit the habits of those precocious birds whose clutches are medium- or large-sized. In such species, when an egg or two fail to hatch within the normal time span of a few hours, the female has long since herded her newly hatched young away from the nest, abandoning the few that did not make their appearance on time. In some woodpeckers, incubation may begin before the set is complete but only after several eggs have been laid. There does not seem to be any particular advantage in this behavioral process, and it is not practiced by very many bird species.

While many passerines practice immediate incubation, the red-billed quelea (*Quelea quelea*) of South Africa demonstrates a striking behavior that is either delayed incubation or at least amazing timing. Gilliard observed that this species nests in colonies with an estimated ten million colonial nests covering an area of three thousand acres. For some unknown reason the egg and incubation cycles are synchronized with the rainy season; this arrangement assures the population of young, hungry nestlings ample food because their diet (as well as that of the parents) consists of seeds and grasses. Many thousands of birds hatch within a few days of each other and the eggshells rain down from their tree nests like snowflakes. Predictably, these hordes of hungry birds can destroy vast quantities of agricultural products. Despite much predation, poisons, explosives, and fire, the population of the quelea does not seem to shrink by any significant degree.

EMBRYO DEVELOPMENT AND HATCHING

Most knowledge of the details of egg development before hatching comes from the study of the chicken. The chicken egg can be opened at any stage to see which embryological structures develop first and how fast each organ grows.

A developing embryo is almost a self-contained world. The energy for its existence lies completely within the shell in the yolk, and the white oxygen necessary for metabolism is drawn through the porous shell and through the extra embryonic membranes. In a fresh egg broken into a frying pan a little white spot on the surface of the yolk can be seen with the naked eye. If the egg is fertile, this spot, called the germ spot, resembles a tiny ring. This is the area that contains the zygote.

Warming a fresh egg to incubation temperature initiates a series of sequential changes that result in the development of the embryo. This remarkable series of events results in a chick ready to hatch just three weeks after the start of incubation. If the egg is opened after several days of incubation, fine red lines representing the start of the circulatory system of blood vessels are visible on the sac that surrounds the yolk. This yolk sac actually consists of two thin membranous structures, sometimes called the extra embryonic membranes. Their functions are respiration and excretion, and they are discussed in some detail in Chapter 2.

If the shell is opened after about three days of incubation, the tiny beginning of the heart can be seen with a magnifying glass. At about the same time—three

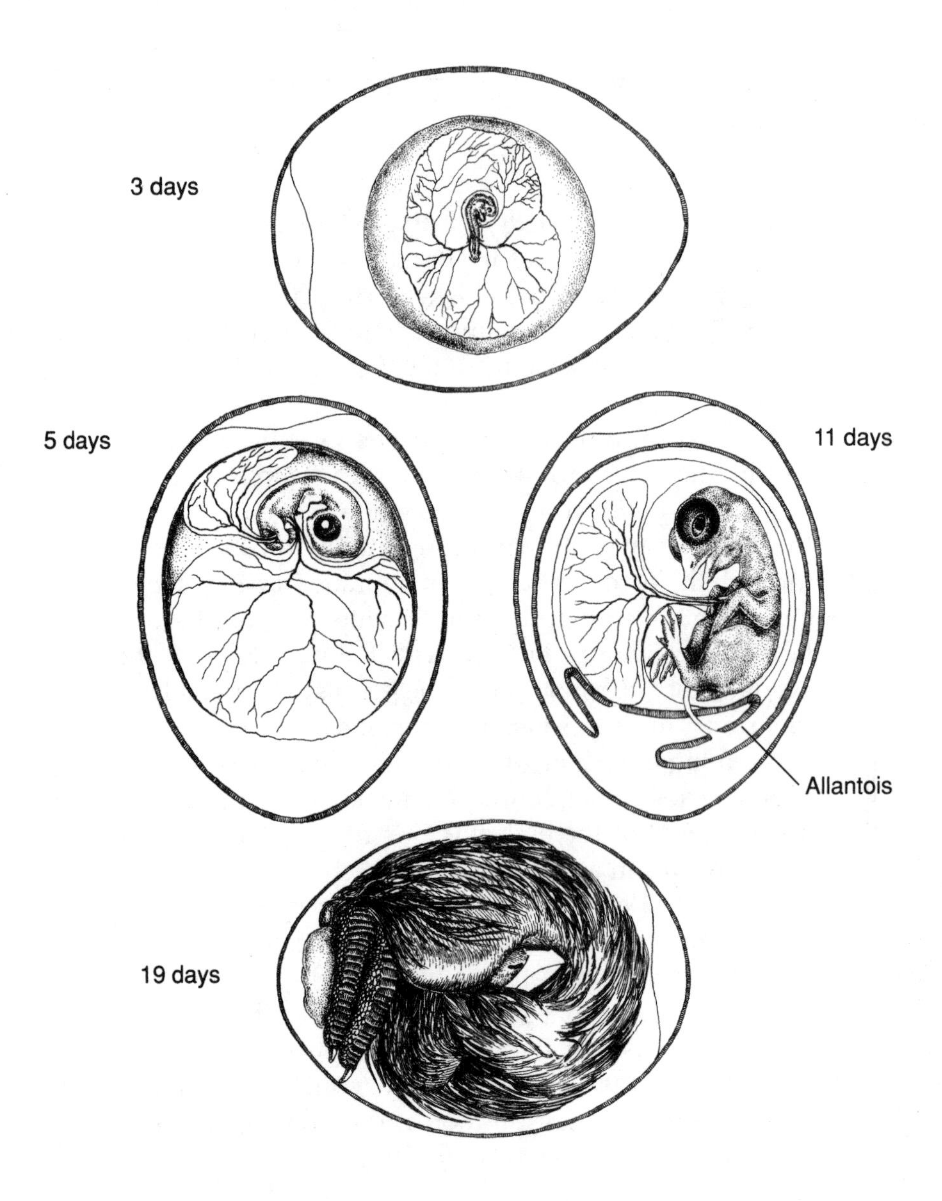

An avian fetus exhibits rapid changes in development. Shown here is an embryo at three, five, eleven, and nineteen days after incubation begins.

days into incubation—the tail bud is also beginning to form. And on the left side of the embryo there is a small, transparent structure that is the beginning of the brain.

At three or four days the amnion is filled with fluid secreted from the developing embryo, which allows the embryo to float. This protects the embryo from shock and makes it possible for the developing embryo to move. After four days of incubation the heart has enlarged, and the brain has started to segment into three sections. Also, the eye is visible as a tiny, dark spot.

At five days the tail, legs, and wing buds are clearly discernable. The allantois membrane has enlarged and is functioning as a depository area for excretory waste that cannot be expelled through the shell. The allantois is a unique structure because it belongs to both respiratory and excretory organ systems. Oxygen needed for the embryo's fast-growing organs is supplied by the allantois; the blood in its vessels is aerated by the air that penetrates the porous eggshell. But the allantois also stores insoluble waste products, which characterizes it as part of the excretory system.

After six days the brain and the eyes are quite prominent. The amnion and allantois are now enlarged and very clearly defined. On the seventh day the beak begins to appear, and the heart is now enclosed in a portion of the yolk sac. On the eighth day the brain is completely enclosed with a membrane, and the embryo lies on its back. Both eyes can be easily seen. Also, the neck is elongated, and the wings and legs are well defined.

On the tenth day the egg tooth forms as a tiny sheath on the top of the bill. Toe digits also begin to appear. Because most of the organs are established by this time, the majority of the growth from now on is simply enlargement. The toes are well formed, toenails and leg scales are visible, and after about two weeks down feathers begin to grow, eventually covering most of the young chicken's body. After fifteen or sixteen days, the developing chick moves its head to a position under the right wing. This brings the beak closer to the shell and makes it easier for the chick to pip the shell when it is time to hatch.

After sixteen days of incubation, the white, or albumen, has become almost completely absorbed, which leaves the yolk as the remaining source of nutrients for the growing chick. Also, around

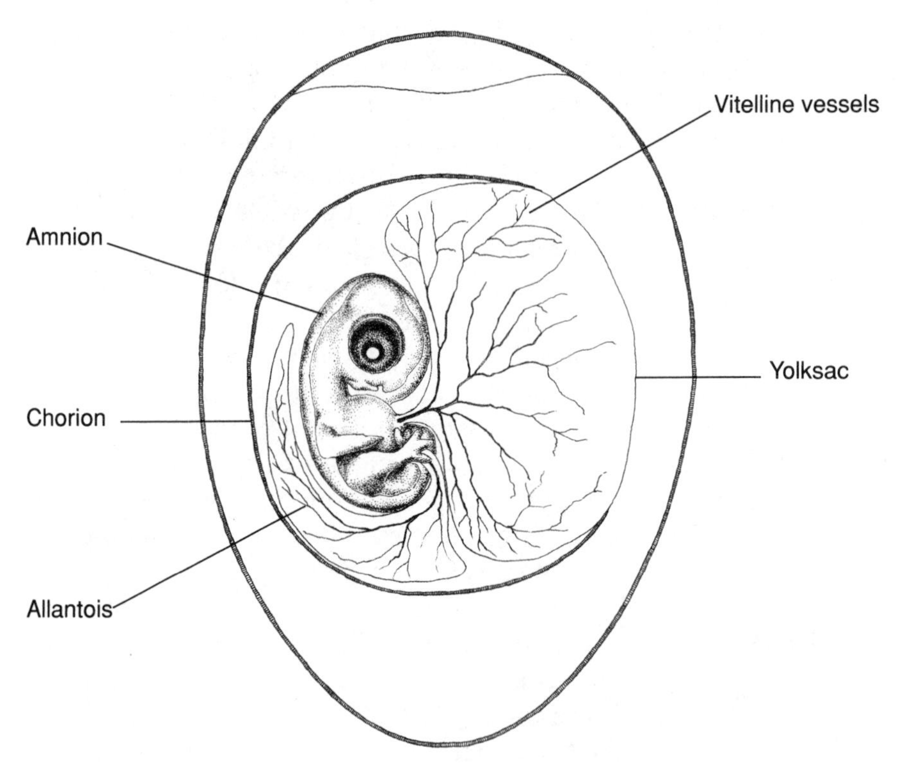

A chicken embryo at seven days, enlarged to show membranes and blood vessels.

the sixteenth or seventeenth day a white waste material—largely insoluble uric acid—can be seen in the allantois. The air cell is getting larger, and the beak under the wing points towards this air cell. In the eighteenth or nineteenth day the chick is preparing to hatch. Yolk material is greatly reduced, and what little is left has been enclosed in the abdominal cavity. This food will supply the newly hatched chick for about the first seventy-two hours of its life outside the shell. Toes, wings, egg tooth, and down feathers are well developed. The beak is ready to pierce the inner shell membrane into the enlarged air cell.

Because most of the egg contents have now been metabolized by the developing embryo and because the air cell is growing larger, there is no longer much inside the shell except the chick and a small amount of waste.

The wing is now over the beak, and the actual process of leaving the shell begins on the twentieth or twenty-first day of incubation. The chick uses the egg tooth on its beak to puncture a small star-shaped hole toward the large end of the egg. When this opening (called a pip) is enlarged slightly, the chick begins to breathe atmospheric air. This marks the initiation of the true breathing process; the developing bird breathes air directly into its lungs. A special hatching muscle (M. *complexus*) develops on the back of the head and upper neck to exert pressure on the shell.

Fisher suggests that the development of the egg tooth parallels the development of the hatching muscle. He notes that the hatching muscle is in excellent position to provide a strong upward thrust of the bill containing the egg tooth. As the hatching muscle develops, an excessive quantity of lymph and water appear around it. Consequently, several biologists have questioned the efficiency of the muscle due to the accumulation of liquid lymph that could hinder muscular action. But the consensus among most ornithologists is that M. *complexus* aids in the hatching process by forcing the egg tooth against the shell. (A few bird species, such as the kiwi, have no egg tooth. The young kiwi's muscular activity alone breaks its shell.)

By pushing with its feet and using its wings as a guide, the chick breaks a ring around the shell. Strong shoving and scratching movements that expand and contract trunk muscles exert pressure, and eventually the shell breaks in two. At hatching, most eggs break approximately at the egg's equator.

Hatching in the domestic chicken is not a quick process; it may take up to eighteen hours of continuous muscular exertion followed by one final heave for the young bird to free itself completely from the shell. It is still wet and weak, but it soon dries out and becomes fluffy. With energy from the nutrients stored in the abdominal yolk it begins to run about.

A number of birds besides the chicken have been studied to determine the details of early embryology. The egg development of some ducks, quail, and others have been worked out in detail, and all show about the same pattern as the barnyard chicken. Differences occur in various birds, due to length of incubation time. The three-week period in chickens compares very well in detail

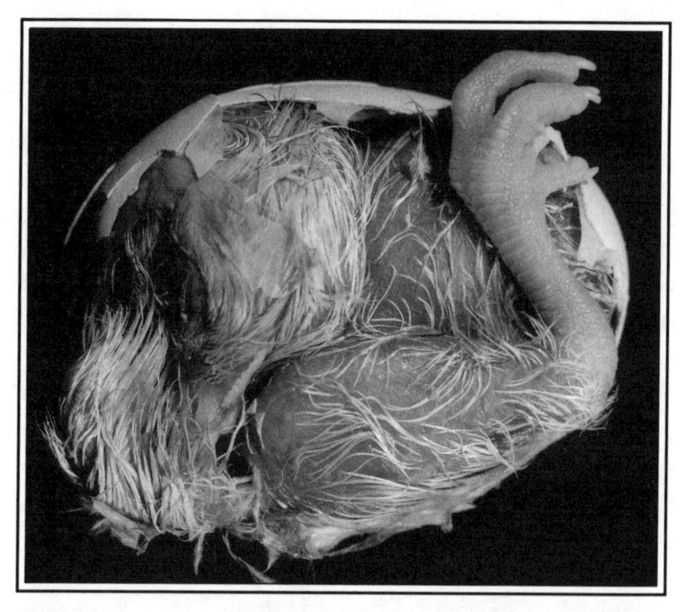

The embryo at sixteen days of incubation. Note the position of the bill under the wing that has been moved down slightly.

The domestic chicken at twenty days of incubation. The egg tooth is visible as a white spot on the tip of the beak.

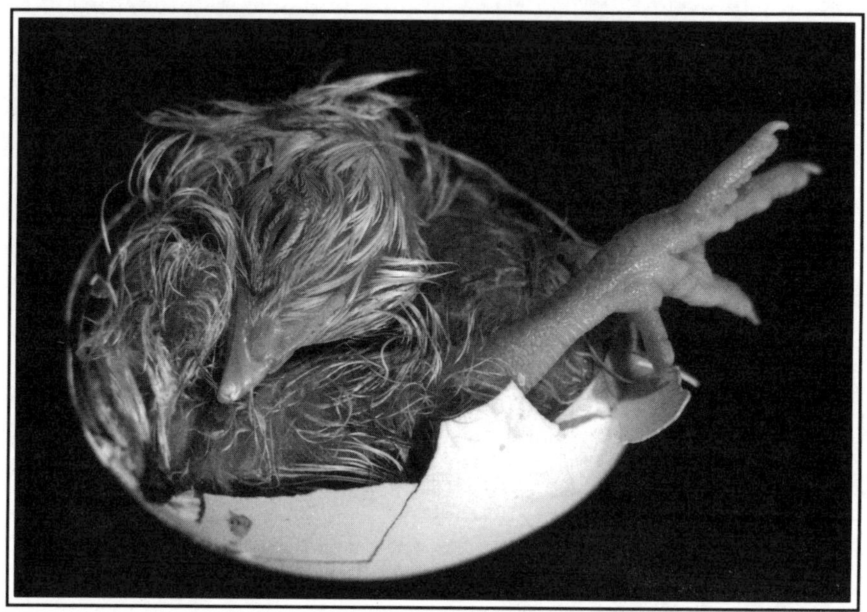

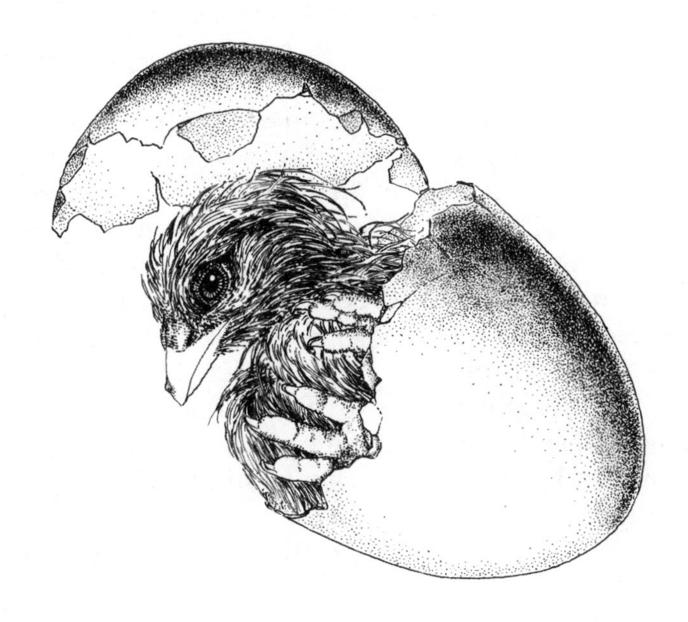

with the embryological processes in the four-week incubation of most ducks. (One could imagine the difficulty of studying the embryological cycle of the tiny hummingbird!)

In the chicken, and presumably many other birds, the chick in the shell has room for some movement, albeit limited. In the last days of incubation, it begins to vocalize. Setting domestic chickens become very excited a day or two before the hatching of their eggs. They may preen, fondle the eggs with their bills, and stay nearby after hearing the young in the shell. Adult male robins have been observed arriving at the nest with offerings of food in the bill as long as twenty-four hours prior to hatching. Their behavior may be a response to vocal sounds and vibrations made by the young robin in its attempts to hatch.

The hatching process demands the expenditure of considerable energy. Double-shelled eggs are not uncommon, especially among some seabirds. Such shells call for an extra expenditure of energy by hatchlings. It is not surprising that many young birds are incapable of breaking through two shells.

At hatching, all that is left behind are fragments of shell and membranes. As a rule, adult precocial species either pay no attention to the shells or eat some of them. But some adult altricial birds will take these shells in their beaks and drop them at a considerable distance from the nest. Presumably, shells strewn about on the ground under the nest would be a telltale sign that young were nearby. Precocial birds, no longer in need of a nest, leave almost immediately after hatching.

Precocial birds begin to develop homoiothermy (endothermy, or warm-bloodedness) before hatching. (Megapodes are completely endothermic at hatching.) Precocial young therefore grow much more slowly than do altricial species because thermal regulation begins early for them. But they are already well developed when they leave the egg, so there is no need for quick growth. The precocial brain is also much more developed at hatching. Precocial birds take but a few hours to become dry; some, like the domestic chicken, are strong enough after drying off to stand, run about, and begin to respond to the voice calls of the mother. Precocity, however, is a relative term. Soon after hatching, seabirds such as gulls and terns are quick to develop the skills to swim and to run in the sand. But they are not able to gather water or pick up food from the ground and are entirely dependent upon adults for a number of weeks. Young terns seem especially helpless, never learning to pick up fresh food from the sand when dropped by an adult. Still, these young terns actively run about at the age of a day or so. But perhaps it is not difficult to understand why downy terns are so dependent, because the species acquires its food—entirely fish—by diving. In this characteristic they are unlike gulls, which have much less selective dietary habits.

Differences in the growth of feathers during incubation can be noted between altricial and precocial birds. Many newly hatched altricial birds have few down feathers, but in precocial birds like the chicken a complete covering of down is well established during incubation. Among marine birds, for example, the precocial cormorants are hatched without any down; in contrast, the newly hatched Leach's storm petrel (*Oceanodroma leucorhoa*), an altricial species, is covered with thick natal down. All newly hatched chicks must be brooded for several weeks to keep them warm, mostly at night or on cold days. During this period one parent usually remains at the nest to protect the young from the cold (and predators), while the other parent brings food to the growing nestling.

Altricial birds emerge from the shell weak, helpless, and exhausted. The eggs of altricial birds may have as little as 50 percent of the amount of yolk found in a typical precocial egg, which is one reason that altricial young are completely dependent on the parents after hatching. These species compensate for their helplessness by growing rapidly. Such growth is possible because the young hatchling doesn't need to expend energy seeking food or establishing its adult thermal regulation process. Newly hatched altricial birds are poikilothermic (exothermic, or cold-blooded) for the first few days. (It has been suggested that this cold-blooded characteristic is a relic of birds' reptilian ancestry.) Since all they have to worry about is growing, strength comes quickly to even the most helpless altricial birds: within a few days many that were very weak at hatching open their mouths, hold their heads erect, move actively in the nest, and in general show great progress in developing.

CARE AND FEEDING OF YOUNG

Most hatchlings need some degree of care from their parents until they fledge. The most important care the adult bird can give is providing food for the young. Instincts are strong enough in the adult that it will start feeding the young in response to movement, physical contact, or other behavior. Joel Welty cites an example of a cardinal perched on the edge of a pool feeding the open mouths of goldfish.

In the most common method of feeding, the parent's beak places the food directly into the beak of the young. In most passerines the adult stands on the rim of the nest to place the food into the mouth of the nestling. The Arctic tern (*Sterna paradisaea*) brings fish to its young from some distance away and hovers over the chick while placing food into its bill. Young herons, on the other hand, must fight over the fish, frog, shrew, mouse, or other small animal carelessly tossed into the nest by the adult.

Frequently the food is offered more or less in its original condition. Young seabirds such as terns, puffins, and auklets, and other fish-eaters such as grebes, herons, and kingfishers, are offered fish within minutes after such food has been collected by the adult. Parent birds capturing insects in the air may place them in the mouth of the nestling while the insects are still alive.

Some species feed regurgitated food to young nestlings, who can take it directly from the bill of the adult. Food may be held in the throat or esophagus of the adult long enough to allow some digestive changes. Young hummingbirds, while in the nest, accept nectar mixed with insects gathered by the adult from conveniently located flowers. The nestling retrieves this food undigested by reaching deep into the esophagus—almost to the stomach—of the adult.

The type of food brought by adults to young, in most cases, indicates the nature of the food that the bird will eat all its life. However, some young hatchlings are not capable of digesting the food on which adults subsist. Young pigeons known as squabs must ingest "pigeon milk," a milk-like substance secreted and shed from the thickened lining of the adult's crop. The release of this food by the adult is apparently stimulated when the nestling inserts its beak into the parent's mouth. Flamingos develop a similar crop fluid. During the breeding season, members of the order Procellariiformes, such as albatrosses and petrels, secrete stomach oils that help preserve the food gathered by adults for the young. They feed the nestlings by regurgitation.

Nestling oil birds from South America are fed a diet of oily fruit. As a result of this rich food, and because they are more or less sedentary in the nest, the nestlings become very fat; the young may weigh twice as much as the adult. (Young oil birds are "harvested" in Venezuela and other areas of northern South America for their fat, used as cooking oil.)

Some colonial nesting birds such as flamingos and certain penguins adopt a peculiar habit of feeding their young. Several days after hatching, the young gather into a large aggregation called a creche. It was once thought that the young were fed by any adult, but L. H. Brown, studying flamingos, noted that parents fed only their own young, which they were able to single out from the individuals forming the creche, by placing food directly into their mouths. This practice does not permit more vigorous young to reap an advantage over less aggressive young. The formation of the creche among penguins may aid in heat conservation; the group of young birds—hundreds or even thousands of the same species—huddled together may retain some warmth in the cold of snow and ice.

Young gallinaceous birds are not fed by adults. The young of these and other species of birds with precocial tendencies are sim-

ply shown food items that they may pick from the ground by themselves. But such food is not directly placed in the mouths of the young. Young precocial birds also detect movements of insects by themselves. Parental feeding is totally absent among brush turkeys and mallee fowl. The young of these strange birds are not fed in the nest, and upon their departure from the nest they are "on their own." Instincts tell them what type of food they must find and eat.

Grebes, coots, and loons are precocial birds that feed on plants, fish, and other foods acquired by diving underwater. The young of these species that have not yet learned to dive must wait on the surface of the lake while the adults fetch the prey, which is then handed over to the young. It is probably significant that these species nest near fresh water, only migrating to salt water after the young know how to dive and secure food for themselves. The more placid waters of a lake make learning to dive easier than it would be in salt water: the currents, waves, and motion of the surf would make learning to dive a longer process and probably a more lethal one. Natural selection resulting from evolutionary changes may have dictated these adaptations in birds that breed in certain areas for greater success, then move to more fruitful, albeit more harsh, environments.

Shortly after hatching, most altricial passerine birds grow rapidly and begin to fill the nest cavity. There is increased competition between nestlings trying to get more than their share of the food brought by the parents and a better position in the nest. Consequently, young altricial birds develop muscle coordination and feeding reflexes rapidly. Only a few days after hatching, vibrations made by the adults arriving with food and landing on the nest edge awaken the young and stimulate them to lift their heads straight into the air, open their mouths, and give voice. The lining of their mouths is highly colored, and large rictal flanges develop and protrude laterally along the edges of the bill. These serve as a target for the adult when the begging nestling thrusts its head in the air with mouth open. The brightly colored bill of many adult seabirds conversely serves as a target for the hungry chick to peck, reminding the adult to serve up a meal.

Certain other physical factors may promote feeding and begging by young birds. For example, air currents set up by adults feeding young swifts in a hollow tree or chimney stimulate the begging response. Young domestic chicks fed on the ground quickly learn to come running when the mother clucks excitedly over a discovered morsel of food.

As a general rule, clutch size in altricial species tends to be smaller when the young are fed in the nest by only one adult. In altricial birds with four or more eggs per clutch, it would be difficult or indeed impossible for one adult to gather enough food for the hungry brood because more and more nourishment becomes necessary to parallel the rapid growth of the young. This is especially true with raptors; despite a small clutch, the young require so much food that both parents must hunt.

Food Acquisition

Because providing nourishment is so important to the survival of the hatchlings, many morphological and structural features in both adults and nestlings reflect extreme adaptations for acquiring food. The long, sharp beak of the kingfisher is perfect for spearing fish, relegating this bird's feet to a minor role in food acquisition. Kingfisher feet are small, vestigial, and used primarily for perching. Similar adaptations have developed in aerial feeders such as swallows, flycatchers, and members of the Caprimulgiformes. In these birds, bills have been reduced in size, but mouths have increased dramatically in spread, making it more feasible and efficient to scoop food from the air. Feet become less useful to these aerial feeders and are consequently highly reduced. Such changes have also occurred in hummingbirds: their long bill for probing into flowers has paralleled a reduction in the size of feet. Hummingbirds can perch but do not walk. Necessarily, they are experts at hovering.

Many seabirds such as auklets, murrelets, puffins, and guillemots dive for fish; their feet and wings are well developed for underwater swimming, and their bills are capable of spearing or otherwise capturing fish. But it is in the raptors that feet and bills become highly associated with food capture. The taloned foot can seize and kill, and the rapier bill can tear flesh into pieces for nestling and personal consumption. Because raptor nestlings require much protein, parent birds must bring the bodies of whole animals to the nest. At first, young raptors are fed bite-sized pieces of tissue that the adults tears from the prey and passes to the young with its beak. Later on, young raptors become capable of tearing the meat and organs from the prey themselves, except for young owls, who are only offered prey that can be swallowed whole.

Success in gathering food and bringing it to nestlings depends in large measure on the wing-lift capabilities of the adult. Wing load is defined as the ratio between body weight and wing area; tail surfaces are not included in wing-load ratios. Birds such as raptors with low wing loads can carry large prey from the kill area to the

nest. Birds with low wing loads are efficient at soaring. But birds with high wing loads need rapid wing beats just to keep aloft; their wing surface is too small for soaring. Species with high wing loads have high landing speeds and cannot fly great distances carrying food of any appreciable weight. Wing load is high in diving birds because a rapid beat is necessary for underwater "flying"; larger wings would be less functional underwater. Swans, coots, large ducks, loons, grebes, and albatrosses have high wing loads. Their body weights are almost too great for the wings to accomplish liftoff. Loons, grebes, and coots must skitter for considerable distances using wings and paddling feet to rise from the water. Birds with long, pointed, narrow wings, such as the albatross, must face into a wind to aid takeoffs; the vulture needs a head wind and a running start to become airborne.

Birds with low wing loads sail gently to a low landing. Low wing loads allow aerial feeders such as swifts, swallows, and martins to maneuver in midair. Kingfishers are also adept at hovering while they search for fish over shallow waters of stream, lake, or bay; hummingbirds hover over flowers. Among raptors, several species are capable of hovering: the kestrel (*Falco sparverius*), the merlin (*F. columbaris*), and one or two owl species. A low ratio also serves another function in many raptors. The low wing load of the owl, for instance, enables it to fly slowly back and forth over a search area looking or listening for prey. Consequently, owls have a slow stall speed at which they can no longer stay airborne. Although they are ground feeders, owls are able to take off with prey in their talons and can easily carry it from the point of capture to the nest to appease the consuming appetites of nestbound young.

Shields notes that early stockmen in California used the concept of wing load in their attempt to control populations of condors with a technique called "penning." Portable pens six feet square and five feet high were baited with a sheep or goat. The birds would settle into the pen and gorge themselves with the "temptingly displayed" feast. When it came time to leave, the bird could neither fly out of the pen, which prevented full extension of its wings, nor jump five feet over the pen because of the additional weight of its meal.

Feeding Frequency

The amount of food brought daily to the nest by parent birds for consumption by the nestlings must be substantial enough to satisfy the needs of fast-developing young. But food-gathering techniques among various species differ greatly, due to diet and nature of prey. Adult raptors might bring two rabbits to the nestlings per day, while passerines offering only one worm or a few insects at one time will need to make many trips. Two small passerine adults feeding a brood of three or four young may each make several hundred round trips between the feeding grounds and the nest per day. Seabirds such as terns may bring to the nestling, in a few trips, fish equal to the weight of the young. Large seabirds such as the albatross feed the chick once daily for about three weeks, after that only twice a week.

Air temperature is also a factor in feeding frequency. On a warm day with twelve or more hours of daylight, two adults may come to the nest with food every few minutes. Throughout the period when the young are being fed, parent birds must be sure that nestlings are provided with sufficient heat, which results in fewer feeding trips on cold days. In many species the adults take turns sitting on the nestlings. An extreme case is represented by the young penguin hatched in the cold of the antarctic. Even though it has a thick natal down, the nestling must be well protected during the feeding period by a fold of skin on the parent's abdomen that can be lowered to cover a chick sitting on the parent's foot. On unfavorable days, nest feeding for most species ceases earlier than it does on warm days. When temperatures are favorable, feeding continues into dusk. On cold, rainy, stormy days, aerial feeders such as swallows, swifts, and members of the Caprimulgiformes may be unable to feed the young until weather conditions permit the capture of insects on the wing. In the meantime, the young live on stored fat. Under adverse conditions of this kind, their metabolism is sharply reduced, and a condition approaching cold-bloodedness keeps the young alive for a matter of days.

Adult nighthawks, poorwills, whippoorwills, and swallows, with their small bills but enormous mouths, can pack their throat and esophagus full of flying insects. The flycatcher—an aerial feeder—can hold in its beak but a few insects and must therefore spend more time in the air searching for insects. However, the wide "scoop" mouth of the swallow and nighthawk catches and holds in one foray a large number of insects. (One bank swallow captured at its nest was found to have a total of twenty-nine insects in its throat and beak.) Thus one feeding excursion by an adult nighthawk produces much more food than a flycatcher or other aerial feeder acquires in many trips. But the nighthawk has only a short feeding period between late afternoon and twilight.

A different condition exists in the rhinocerous auklet, which captures food in water. The auklet may pack its beak with several fish, each measuring many inches in length. The small fish, held crosswise in the auklet's beak, look like a large, silvery moustache; as many as twenty-two have been counted after one foray. The auklet spends its daylight hours in the burrow and feeds its young only at night; its inordinate skills in capturing fish make fewer forays necessary. The adult birds, bringing fish to the developing young in the burrow, alight in the grasses several feet from the burrow entrance, and with a crouching run disappear into the proper burrow. At such times these birds may be confused by the glare of a flashlight, which makes them easy to capture.

Postnest Feeding

The time between hatching and departure from the nest is a critical period in the life of the new bird. Several general patterns are obvious in this growth period in the nest. Altricial young that are fed directly by the parents grow rapidly. Self-fed precocial birds grow slowly. As a rule, large species grow much more slowly than small species. In many cases, the young altricial birds will become heavier than the adult while still in the nest. This phenomenon is especially common for many seabirds, pigeons, tube-nosed birds, and oil birds.

The young of many altricial species beg for food from adults for some time after they leave the nest. This post-nest feeding in such birds as robins, sparrows, and grosbeaks may extend over a period of days. Hungry young sit on the ground, extend and wave their wings, and cheep shrilly when the adults make their appearance with food offerings.

Behavior patterns that develop when adult birds feed young on the ground are similar to those that develop during late incubation. If an enemy approaches when young birds are on the ground, the adult bird is likely to attempt to draw the invader away by feigning an injury. This behavior is strikingly similar to that observed in adults when eggs are hard set. The display often includes a combination of plaintive cries, alarm notes, or a struggle faking some physical ailment. The "broken wing act" is common among gallinaceous birds, some ducks, and some shore birds and is also observed in the long-eared owl. It is also a frequent practice for many adult birds to "dive-bomb" an intruder in a nesting site when young are present. Dive-bomb behavior is widespread throughout most orders of birds and is commonly observed in eagles, hawks, terns, some gulls, and the olive-sided flycatcher (*Contopus borealis*).

Toward the end of the nestling period, feeding trips are reduced or stopped completely. This may be due to fatigue in the adult, but it may also be part of the normal behavior pattern that encourages the hungry young to strike out on its own to find something to eat. In some species, adults may even begin their annual migration some days before the young leave the nest.

I watched the typical nesting process of a pair of western robins (*Turdus migratorius*) outside my breakfast window at a distance of about three feet. I observed two nesting cycles—1983 and 1984—in detail. The spring of 1983 was typical. The nest was completed and three eggs laid by May 12, when incubation started. A fourth egg was laid one day later. On May 22 the male made frequent trips with offerings of worms, but hatching did not occur until May 23. Like the hen chicken that becomes very active and agitated the day before her chicks hatch, the robin perhaps detected some movement and sound within the egg. Feeding began on May 24 and continued until June 7, when all four young left the nest together. Thereafter, the male robin appeared at the empty nest several times with a mouthful of worms. Thus, feeding stretched over a period of fifteen days with a feeding day that lasted approximately twelve hours.

Other Forms of Care

In addition to providing sufficient food of the right kind to nestlings and providing warmth for developing young, adults may make unusual efforts to provide some temperature reduction for nestlings hatched in desert regions devoid of shade. Adult sand grouse supply drinking water to nestlings when there is insufficient water in the food; they wade into lake or river and soak up water in special absorbent feathers on the abdominal region (Cade and MacLean). Koenig considers prey dunking by Brewer's blackbirds (*Euphagus cyanocephalus*) a novel source of water for nestlings. He observed a large proportion of blackbirds in the Hastings Reservation in Monterey, California, coming to a pond to dunk grasshoppers before feeding them to the young in the nest. He concluded that the water clinging to the body of the grasshopper is an important source of liquid to the nestlings. Koenig suggests that other species of birds nesting in dry environments may dunk prey before feeding nestlings.

Another form of care for the young is observed in certain raptors when brooding or feeding. Adult ospreys, for example, commonly clench their talons into fists to make sure the sharp claws do not injure the young.

Proper nest sanitation is essential in most species for the healthy development of young still in the nest. But a few species perform no sanitation procedures. The nests of most pigeons become so completely unsanitary that they become culture media for bacteria, molds, and many varieties of insects and other invertebrates. For many species, nest sanitation simply means somehow disposing of nestling feces. In most cases, the excreta of the young are enveloped in a thick mucous membrane forming a small, oval sac. It is a common practice, especially among passerines, for the adult to eat this fecal sac. This behavior of adult ingestion is economical; the rapid movement of food through the nestling's alimentary canal leaves some nourishment in the sac available to the adult. Additionally, the adult is often too busy to hunt for itself. In many species, the fecal sacs are eaten by the adults only in the first few days after feeding begins. Thereafter the sacs are carried off and dropped at some random distance from the nest.

Studies with various species of birds seem to indicate that parent birds dispose of sacs in scattered areas rather than drop them in a single disposal area. Fecal sacs concentrated in a particular area would probably indicate to a predator that nestling birds are in the vicinity. Such birds as swallows typically drop the sac over water. The lyrebird of Australia very carefully submerges the sac underwater. Ornithologists have been quick to assume that removal of fecal sacs is primarily an aid to sanitation. But Weatherhead questions this assumption and wonders why adults carry the sac considerable distances from the nest. He suggests that deposition of the sacs away from the nest helps preserve privacy of the nest. This hypothesis gets some credence from the fact that certain hole-nesters concealed in the nest cavity do not produce sacs or they eat the sacs. Furthermore, raptors and some seabirds that are not subject to predation, and thus do not need to worry about excreta being a telltale sign to a roving predator, do not remove their feces.

For the first few days of feeding many species remove the sac from the bottom of the nest. Toward the end of the nest phase, if no sac is present after feeding, the parent will often shake or gently peck the young, thus encouraging it to stand up in the nest and expel the sac so that the adult can pick it directly from the cloaca.

Although most Passeriformes practice nest sanitation by disposal of fecal sacs, a few species quite often become careless, sometimes allowing nests to become highly unsanitary for a time prior to hatching. The pine siskin (*Carduelis pinus*) is one of these careless species, and some years an "epidemic" of unsanitary habits is highly visible in a number of siskin nests. E. A. Kitchin describes a day in March of 1927 when he visited a "convention" of siskins in fir woods adjacent to open prairies in the northwestern United States. Every grove in an area of five square miles contained hundreds of birds. One week later when he again visited the area, the woods were empty and silent. What had become of the birds that had eggs? Then he discovered "one of the most remarkable stunts ever performed by a living bird." All but a few of the many nests he examined were "so befouled with excreta that the cup was filled to the brim, and in every such case the unhatched eggs were underneath!" All of the birds had deserted, but they "buried and befouled their treasures before leaving." Not one egg was broken. When cleaned, those eggs found their way into oological collections. Kitchin did not know why the birds left so abruptly, but he cited old records going back as far as 1867 describing the behavior of siskins in Pennsylvania in ways "almost identical with ours."

FLEDGING

LEAVING THE NEST

Webster's dictionary says that to fledge is "to rear or care for a bird until its plumage is developed so that it can fly." A fledgling is "a young bird just fledged"—that usually means one that has left the nest. For instance, young swifts of the genus *Apus* are independent and able to fly when they leave the nest. But other species of birds demonstrate different fledging patterns. Some young birds can fly but are completely dependent upon the parents for food: young robins, swallows, grosbeaks, and blackbirds fly from the nest but are fed on the ground by parents until they are old enough to seek out their own food. Ground nesters such as gulls and terns leave the nest soon after hatching but depend on parents completely for food for several weeks. These young seabirds are active on the ground and may even swim well but are unable to feed themselves or fly. Raptor young may utilize the nest throughout the fledging period, which can be as long as several months for some species. The fledging period is so long for the albatross (up to eleven months) that this species usually has time for a nesting cycle only every other year.

Fledging is really a much more relevant term for altricial birds than it is for precocial species since precocial young need very little care after hatching and become independent quickly. If the above definitions are used literally, some precocial birds pass through a very brief fledging period or none at all. If the swift mentioned above can fly once it leaves the nest it does not—by definition—have a fledging period. So, too, the brush turkey does not have a true fledging period. This bird flies when it pushes up through the nest mound to emerge into the outer world.

Survival of the Gene Pool

A population, also known as the gene pool, must be kept large enough to assure the continuance of a species. By no means does the population of every species stay at a minimum or optimum level, and, consequently, reduction in numbers of a species may fluctuate from season to season. If, over a critical period of time, the pool shrinks below the minimum number for growth or replacement of the species, extinction may result.

A classic example of this principle is demonstrated in the tragic extinction of the passenger pigeon (*Ectopistes migratorius*). This species is thought to hold the record for the largest bird population ever known to man. In spite of its small clutch—one egg (sometimes a second) hatched successfully in one season—its reproductive cycle was so successful that over a period of years, millions of birds migrated from South America to the eastern United States annually. Changes—some due to natural forces, some due to man—reduced the population of the passenger pigeon to a few hundred thousand. Some ornithologists and others became alarmed and strove to enact laws regarding the killing of the species; others scoffed, saying that there were still enough birds to continue the kill without endangering the entire population. But with reproduction at such a low level, only one or two squabs a year added to the gene pool, the normal processes in the reproductive cycle could not overcome the population handicap.

The California condor represents another obvious example. Its every-other-year chick did not add enough individuals to the population to overcome the adverse environment created by shrinking habitat and other incursions of man. At the present time no California condors are left in the wild; the twenty-five remaining members of the species are all confined to zoos or aviaries.

The histories of extinctions in most birds seem to point to very large populations followed by a decline in numbers to a population size so small that recovery was impossible. But where certain

species can adapt to changes or can easily adjust their reproductive cycles to a variety of conditions, maximum or minimum populations continue for generation after generation. Such versatile species as the European starling, house sparrow, and weaver finch (*Quelea* sp.), which have adjusted their reproductive cycles to encompass a great variety of habitats and conditions, have huge populations that continue to grow, despite man's special efforts to reduce or at least prevent an increase in their populations. It follows, then, that birds that can adjust to a variety of conditions in their reproductive cycles are successful in perpetuating their species. Those that are unable to adjust to changes are reduced to extinction.

All the factors that govern the steps in the reproductive cycle, from selection of territory to fledging, have brought about adaptations in the behaviors and habits of species that assure success and continuity of the species. At the conclusion of the reproductive cycle, whether in a great colony of identical individuals or in a region shared by many species, young that are successfully raised past fledging become a part of the reproductive pool. Barring catastrophic events, they ensure continuation of the species.

A COMPARISON OF VARIOUS CHARACTERISTICS IN PRECOCIAL AND ALTRICIAL BIRDS

The following table shows the characteristics that differentiate precocial birds from altricial birds. Not all birds of one type will exhibit all the characteristics listed; for instance, the precocial kiwi has a clutch of only one or two. Nevertheless, these features of the reproductive cycle serve as a general guide for identifying a bird as precocial or altricial.

PRECOCIAL	ALTRICIAL
large eggs	small eggs
much yolk	minimal yolk
large clutch size	small clutch size
delayed incubation	immediate incubation
long incubation period	short incubation period
short nest period	long nest period
large chick	small chick
much down	little or no down
great attrition of young	less attrition of young
short poikilothermic period	long poikilothermic period
partially dependent on adults	totally dependent on adults
slow growth rate	rapid growth rate
do not remove shells from nest site	remove shells from nest site

GLOSSARY

Allantois — Embryonic membrane used for excretion and respiration.

Allopatric — Refers to two species with contiguous breeding ranges that are separate geographically. (See Sympatric.)

Altricial — Having the young hatched in a very immature and helpless condition, usually blind and naked, so as to require care for some time.

Brood Parasitism — The practice in certain bird species of depositing their eggs in the nest of another species.

Broodiness — A behavior pattern that develops in the female bird for covering the eggs and young.

Chorion — The third of the embryonic membranes, which eventually covers the dorsal surface of the embryo.

Egg Tooth — Temporary horny covering of the tip of the bill in newly hatched birds.

Endothermy — See Homoiothermy.

Exothermy — See Poikilothermy.

Follicle — A cell in the ovary that gives rise to the ovum.

Galliformes — An order of heavy-bodied, largely terrestrial birds including the pheasant, turkeys, grouse, and the common domestic fowl.

Gallinaceous — Of or relating to birds of the order Galliformes.

Homoiothermy — A body temperature remaining constant regardless of environmental temperature. Commonly called warm-bloodedness.

Jurassic — A period of geologic history lasting about 50 million years, beginning about 180 million years ago.

Megapode — Large-footed gallinaceous birds of Australia. Their eggs are incubated in large earthen mounds.

Monogamy — A pair bond formed by one male and one female for one season, several seasons, or for life.

Nearctic — North America, Greenland, and Iceland.

Neotropical — Central and South America.

Nidology — The study of bird nests and nest building.

Oology — The science of the study of eggs.

Oviparous — Producing eggs that are developed and hatched outside the maternal body.

Ovoviviparous — Producing eggs that develop within the maternal body and hatch within or immediately after extrusion from the parent.

Palearctic — Asia and Europe north of the Sahara Desert.

Passeriformes — The largest order of birds, which includes more than half of all living birds and consists chiefly of altricial songbirds of perching habits.

Passerine	Of or relating to birds of the order Passeriformes.
Peristalsis	The contraction of circular muscles in tubular structures.
Piciformes	The order of birds comprising woodpeckers, flickers, sapsuckers, and toucans.
Poikilothermy	Body temperature varying with that of the environment. Commonly called cold-bloodedness.
Polyandry	The mating of one female with more than one male in a season. In addition, the male tends to the nesting, feeding, incubation, and care of young.
Polygamy	The mating of one individual with more than one individual of the opposite sex in a season. (See Polyandry and Polygyny.)
Polygyny	The mating of one male with more than one female in a season.
Precocial	Capable of a high degree of independent activity from birth. Precocial birds hatch with a covering of down, their eyes open, and are ready to walk or swim almost immediately.)
Procellariiformes	The order of birds comprising the tube-nosed swimmers, i.e. albatrosses and petrels.
Promiscuity	Copulation with any individual of the opposite sex where no bonding occurs.
Raptor	A predatory bird such as eagle or owl.
Rictal Flange	A colored fold of skin outlining the beak in a newly hatched altricial bird.
Ruptive Marks	Patterns of color on certain birds that tend to break up the outline of the bird and thus aid in concealment, as on the killdeer and harlequin duck.
Siblicide	The killing of a nestmate.
Sympatric	Refers to two species whose breeding ranges merge and overlap. (See Allopatric.)
Syrinx	The voice box in birds located at the base of the neck at the bifurcation of the trachea.
Taxonomic Group	A number of organisms classified together by their anatomical and morphological similarities.
Triassic	A period in geologic history lasting about 50 million years, beginning about 220 million years ago.
Vas deferens	In male birds, the tube leading from the testes to the cloaca.
Viviparous	Producing young that are born alive and have developed through the embryonic stage deriving their food from a placenta in the manner of most mammals. No birds are viviparous.
Wattle	A colored structure largely of connective tissue, found around the head and neck of certain birds and used for display.
Zygodactyl	Having the toes arranged two in front and two behind. Found in certain birds such as the Piciformes.
Zygote	Embryo that results from the union of sperm and egg.

BIBLIOGRAPHY

Alcorn, Gordon D. *Owls: An Introduction for the Amateur Naturalist*. New York: Prentice Hall Press, 1986.

Alcorn, Gordon D. and Garrett Eddy. "The Nesting of the Oregon Leach Petrel on the Quillayute Needles." *The Murrelet* 35:3 (1954): 46-47.

Austin, O. L. *Birds of the World*. New York: Golden Press, 1983.

Barnes, R. Magoon. *American Oologists' Exchange Price List of North American Birds' Eggs*. Lacon, Illinois: published by author, 1922.

Barrett, C. L. "The Origin and Development of Parasitical Habits in the *Culculidae*." *The Emu* 6 (1906-7): 55-60.

Bowles, J. H. "Nesting of the Western Robin." *The Murrelet* 8:3 (1922): 74.

Brown, L. H. "The Breeding Biology of the Greater Flamingo, *Phoenicopterus ruber*, at Lake Elmenteita, Kenya Colony." *Ibis* 100 (1958): 388-420.

Cade, T. J. and G. L. MacLean. "Transport of Water by Adult Sand Grouse to Their Young." *Condor* 69:4 (1967): 323-343.

Clark, George A., Jr. "Life Histories and the Evolution of Megapodes." *The Living Bird* Volume 3. Ithaca: The Laboratory of Ornithology, Cornell University, 1964.

Cooper, William T. and Joseph M. Forshaw. *The Birds of Paradise and Bower Birds*. Boston: David R. Godine, 1979.

Coues, Elliott. *Key to North American Birds*, 2d ed. Boston: Estes and Lauriat, 1884.

Dawson, W. L. and J. H. Bowles. *The Birds of Washington*. Seattle: Occidental Publishing Company, 1909.

DeBeer, Sir Gavin. *Archaeopteryx lithographica*. Trustees of the British Natural History Museum, 1954.

Feduccia, Alan. *The Age of Birds*. Cambridge: Harvard University Press, 1980.

Fisher, H. I. "The Hatching Muscle in the Chick." *Auk* 75 (1958): 391-399.

Friedmann, H. *The Cowbirds: A Study in the Biology of Social Parasitism*. Springfield, Illinois: Charles C. Thomas, 1929.

———. "Host Relations of the Parastitc Cowbirds." *United States National Museum Bulletin* 233 (1963): 1-276.

———. "Additional Data on the Host Relations of the Parasitic Cowbirds." *Smithsonian Miscellaneous Collections* 149 (1966): 1-12.

Friedmann, Herbert, Lloyd F. Kiff, and Stephen I. Rothstein. "Further Information on the Host Relations of the Parasitic Cowbirds." *Auk* 88 (1971): 239-255.

Frith, H. J. "Temperature Regulation in the Nesting Mounds of the Mallee Fowl, *Leipoa ocellata* Gould." *Wildlife Research* 1 (1956): 79-95.

———. "Experiments on the Control of Temperature in the Mound of the Mallee Fowl, *Leipoa ocellata* (Megapodidae)." *Wildlife Research* 2 (1957): 101-110.

Gilliard, E. Thomas. *Living Birds of the World*. New York: Doubleday, 1958.

Grant, Gilbert S. "Avian Incubation: Egg Temperature, Nest Humidity, and Behavioral Thermoregulation in a Hot Environment." *Ornithological Monographs* No. 30, A. O. U., Washington, 1982.

Greenwalt, Crawford H. *Hummingbirds.* Garden City, New York: Doubleday, 1960.

Hamilton, W. J. and G. H. Orians. "The Evolution of Brood Parasitism in Altricial Birds." *Condor* 67 (1965): 361-382.

Harrison, C. J. O. *Bird Families of the World.* New York: Harry N. Abrams, 1978.

Ingram, C. "Cannibalism by Nesting Short-eared Owls." *Auk* 79 (1962): 715.

Johns, J. E. "Field Studies of Wilson's Phalarope." *Auk* 86 (1969): 660-670.

Kinsky, F. C. "The Consistent Presence of Paired Ovaries in the Kiwi (*Apteryx*) With Some Discussion of this Condition in Other Birds." *Journal Fur Ornithologie* 112:3 (1971): 334-357.

Kitchin, E. A. *Birds of the Olympic Peninsula.* The Olympic Stationers, 1949.

Koenig, Walter D. "Dunking of Prey by Brewer's Blackbirds: A Novel Source of Water for Nestlings." *Condor* 87:3 (1985): 444-445.

Mayfield, H. F. "Chance Distribution of Cowbird Eggs." *Condor* 67 (1965): 257-263.

———. "Census of Kirtland's Warbler in 1972." *Auk* 90:3 (1973): 684-685.

Phillips, Charles L. in Bent, Arthur C. "Life Histories of North American Woodpeckers." *United States National Museum Bulletin* 174 (1939): 272.

Ripley, S. D. "Strange Courtship of Birds of Paradise." *National Geographic* 97 (1950): 247-278.

Shields, A. M. "Nesting of the California Vulture." *Nidologist* 2 (1895): 148-150.

Smith, L. H. *The Lyrebird.* Melbourne: Lansdowne Press, 1968.

Thomson, Charles F. and Bradley M. Gottfried. "How Do Cowbirds Find and Select Nests to Parasitize." *The Wilson Bulletin* 88:4 (1976): 673-675.

Tinbergen, N. *The Herring Gull's World: A Story of the Social Behavior of Birds.* London: Collins, 1953.

Van Tyne, Josselyn and Andrew J. Berger. *Fundamentals of Ornithology.* New York: John Wiley & Sons, 1959.

Weatherhead, Patrick J. "Fecal Sac Removal by Tree Swallows: The Cost of Cleanliness." *Condor* 86:2 (1984): 187-191.

Welty, Joel Carl. *The Life of Birds.* New York: Saunders College Publishing, 1982.

Wilson, Edward O. *Sociobiology, the New Synthesis.* Cambridge: Harvard University Press, Belknap Press, 1975.

INDEX

* Italic page numbers refer to illustrations.